S0-CFD-667

Embracing Mind

embracing
Mind

THE COMMON GROUND OF SCIENCE AND SPIRITUALITY

B. Alan Wallace
and Brian Hodel

SHAMBHALA
BOSTON AND LONDON
2008

Shambhala Publications, Inc.
Horticultural Hall
300 Massachusetts Avenue
Boston, Massachusetts 02115
www.shambhala.com

© 2008 by B. Alan Wallace and Brian Hodel

All rights reserved. No part of this book may be reproduced in any form
or by any means, electronic or mechanical, including photocopying,
recording, or by any information storage and retrieval system, without
permission in writing from the publisher.

9 8 7 6 5 4 3 2 1

First Paperback Edition
Printed in Canada

⊗ This edition is printed on acid-free paper that meets
the American National Standards Institute Z39.48 Standard.
♻ This book was printed on 100% postconsumer recycled paper.
For more information please visit us at www.shambhala.com.
Distributed in the United States by Random House, Inc.,
and in Canada by Random House of Canada Ltd

Designed by Gopa & Ted2, Inc.

The Library of Congress catalogues the hardcover edition of this
book as follows:
Wallace, B. Alan.
Embracing mind: the common ground of science and spirituality/
B. Alan Wallace and Brian Hodel.
p. cm.
Includes bibliographical references and index.
ISBN 978-1-59030-482-2 (hardcover: alk. paper)
ISBN 978-1-59030-683-3 (paperback)
1. Science—Philosophy. 2. Religion and science. I. Hodel, Brian. II. Title.
Q175.W2618 2008
261.5'5—dc22
2007033698

Contents

Preface

I AM ONE of the many fortunate people who was inspired at an early age by a teacher who gave direction to the rest of their life. In my case, it was Sally Vogel, my science and mathematics teacher when I was thirteen years old, growing up in southern California. Sally's influence on my goals and aspirations did not come through the traditional classroom, but through her love for nature and dedication to caring for the environment. On weekends she invited her students on nature hikes and during these outings she shared her infectious enthusiasm for understanding and appreciating our natural environment. She was a passionate advocate of preserving it and inspired me as a young adolescent to dedicate my life to those same ideals.

Sally is a humanist and a naturalist, with no religious affiliation, but my uncle, Dave Needham, is a devout Christian, who expressed a similar love for the wilderness. Each summer, he would lead a group of teenage boys from the church where he served as pastor on weeklong backpacking trips into the Sierra Nevada Mountains. Each day we would hike for about ten miles among the towering peaks and aquamarine lakes of the majestic High Sierras. In the evenings around the campfire, Dave would read passages from the New Testament and lead us in discussions of their implications for our lives. His life was devoted to the pursuit of emulating the love of Jesus and of cherishing God's creation.

Throughout high school, I studied with the thought of pursuing a career in ecology, and drew great inspiration from the writings of Henry David Thoreau. One passage from *Walden* made a particularly deep impression: "I went to the woods because I wished to live deliberately, to front only the essential facts of life, and see if I could not learn what it had to teach, and not, when I came to die, discover that I had not lived."[1] My other literary hero then was John Muir, with his vision of nature as one interconnected system that originated in the same divine source. When this was properly understood, it could offer salvation to a society that had lost touch with its divine origins.

My youth was characterized by two major influences: growing up in a religious family with a Christian theologian for a father and this wish from an early age to dedicate my life to science. Yet, the more I learned about the Christian worldview and that of modern materialist science, the more dissonance there seemed between them. The first provides culture with a lofty set of ethical ideals and a vision of Creation imbued with God's will and guidance, while the latter presents a massive body of knowledge of a universe governed by chance and necessity. During two years as an undergraduate at the University of California at San Diego, I sought to integrate the deeply spiritual and the rigorously scientific. However, I found no one in the church or on campus who could help me realize this goal.

These final years of the 1960s were for me a time of rapidly growing discontent and alienation. The United States was waging an unnecessary, tragic war in Vietnam, and as an active member of the environmental movement, I found myself growing more and more outraged at the abuses of the ecosphere. My own mind was becoming similarly polarized into viewing fellow environmentalists on the side of goodness and sanity and everyone else on the side of greed and delusion. Gradually, I recognized that my own attitude of self-righteousness and hostility were part of the problem, not part of the solution.

It was time for a change, and after this second year at the university, I spent a summer hitchhiking around Europe, followed by a year at the University of Göttingen in Germany. During that summer of 1970, while staying at a youth hostel in the Swiss Alps, I came across a book that changed the course of my life. It was W. Y. Evans-Wentz's *The Tibetan Book of the Great Liberation: Or the Method of Realizing Nirvana Through Knowing the Mind*, with an introduction by Carl Jung. Here for the first time I came upon a Buddhist vision of reality and a way to explore it experientially that held the promise of satisfying this yearning to integrate a scientific and spiritual way of engaging with the world.

Through the summer, I continued slowly reading this book, until I arrived in Bergen, Norway. At this point I felt an urgent need to find a "wise old man," as I wrote in my journal, who could give me the guidance I sought. The next morning, after being dropped off several hours outside of town on the long road to Oslo, I was picked up by a wizened little old man driving a black VW. We immediately fell into conversation and I quickly learned to my astonishment and delight that he was a Buddhist monk who had lived for years with Tibetans in Nepal. He gave me just the guidance I was seeking, and I continued my journey with a new confidence that reality was rising up to meet me in my quest for a meaningful, integrated life.

Back at Göttingen, I quickly learned that the university offered no courses on ecology and a rather dry array of courses on philosophy. It did have a fine department for Indic studies, however, and on its faculty was a Tibetan lama named Dzongtse Rinpoche. Although I knew very little about Tibetan Buddhism at the time—there were only about a half dozen decent books on the subject—I was powerfully drawn to it, and it was obvious that if I was to devote my life to exploring it, I had to learn the Tibetan language. So I dropped all other classes for which I'd registered, and became Dzongtse Rinpoche's only language student.

In the meantime, I was considering the diversity of the world's religions. There was no question that their doctrines differ in many significant ways, and many of them claim to be uniquely true and to provide "the only way" to salvation or liberation. But might there be an underlying unity, or at least convergence, in the experiential insights of the great contemplatives throughout history, East and West? As the months passed, I read everything I could find on the contemplative traditions of the world and eventually I encountered Aldous Huxley's remarkable work, *The Perennial Philosophy*. Its central premise is that the mystical traditions of the world converge upon a common reality that transcends words and thoughts: this is the ultimate ground of being. It struck me that if there was such a convergence, the truths they revealed must be the most important ones human beings could discover. I was willing to devote the rest of my life to exploring them.

As my year in Göttingen drew to a close, I had a clear sense of the way ahead: I had decided to go to Asia, become a Buddhist monk, and devote the next ten years to studying and practicing Tibetan Buddhism. Dzongtse Rinpoche, however, encouraged me first to spend the summer in a Tibetan Buddhist monastery in Switzerland. While I was there, a bulletin arrived from Dharamsala, India during the summer of 1971, announcing that the Library of Tibetan Works and Archives was soon to open, and under the guidance of the Dalai Lama, classes on Buddhism were to be offered for Westerners. After selling or giving away everything I couldn't carry on my back, I bought a one-way ticket to India, and for the next four years immersed myself in the Tibetan language, Buddhism, and culture. And, true to that earlier aspiration, I became a Buddhist monk, ordained by the Dalai Lama himself.

While my original goal had been to integrate a spiritual and scientific view of the world, these years spent among the Tibetans had the effect of adding a third element. The Tibetan Buddhist paradigm has its own internal coherence yet it is radically different from both the Judeo-Christian and Greco-Roman heritages that have defined Western civilization. In particular, in its pursuit of objective truth, science has split the world into an objective reality that exists independently of experience and a subjective world of mental experiences that are thought to arise from physical processes. On the other hand, Buddhism is concerned with understanding the *world of experience,* which doesn't exist without consciousness. Consequently, in the West, scientific study of the mind did not begin until more than three hundred years after Copernicus, and for the most part it has been studied by way of physical processes, namely behavior and neural functions. In Buddhism, on the other hand, the experiential study of the mind and consciousness have always been of central importance, and the focus has been on mental phenomena themselves, and not just their physical causes and effects.

After four years in Dharamsala, the Dalai Lama encouraged me to accompany my teacher, Geshe Rabten, back to the monastery in Switzerland. So for the next five years, I continued my monastic training under his guidance and that of many other Tibetan lamas for whom I served as interpreter. I also began teaching Buddhist philosophy and meditation, but felt a growing need to assimilate what I had learned over the years by applying myself wholeheartedly to meditation. When the Dalai Lama encouraged me to return to Dharamsala, where he offered to train me in meditation himself, I leapt at the extraordinary opportunity and in the spring of 1980 returned to India. I soon moved into a rustic meditation hut high in the mountains above Dharamsala and stayed meditating there until my Indian visa expired and I had to leave the country. For the next three years, I continued in a series of solitary meditation retreats in India and the United States until I was ready to begin the long process of integrating what I had learned through years of Buddhist study and practice with that earlier scientific and religious background.

I knew this was possible because although science is the dominant paradigm of knowledge for many people throughout the world, Buddhism, too, presents itself as a tradition based on empirical knowledge, not simply religious faith. Science, however, has always marginalized the role of the mind in the natural world while Buddhism has emphasized experiential investigation and transformation of the mind. I thought that the great strength of science lies in its exploration of objective, physical, quantifiable phenomena, and the branch of science that best represents this mode of inquiry is physics. So, after fourteen years away from Western academia, I recommenced my undergraduate education at Amherst College to study physics along with the history and philosophy of science, and to relate these to Buddhist theory and practice.

As a mentor in physics, I had the privilege of enlisting the Amherst physicist, Arthur Zajonc. Robert Thurman taught Sanskrit, and other gifted scholars guided me in the study of the history and metaphysical foundations of modern physics. After a first semester I was granted the status of "Independent Scholar," which gave virtually complete freedom to pursue studies on and off campus. While continuing to take courses in physics and mathematics, I focused my scientific research on the "zero-point energy" of empty space. Philosophical studies were focused on the writings of the great pioneers of twentieth-century physics, including Albert Einstein, Niels Bohr, Werner Heisenberg, Erwin Schrödinger, and Louis de Broglie.

As I explored this revolutionary movement in the scientific conception of nature, it became clear that physics, the most mature of the natural sciences, was undermining the classical assumption that science can describe an absolutely objective reality that exists independently of human experience and concepts. As Heisenberg declared, "what we observe is not nature in itself but nature exposed to our method of questioning."[2] This inspired a comparison between the philosophical foundations of modern physics and the Buddhist Middle Way philosophy (Madhyamaka), which declares that all objective and subjective phenomena arise as dependently related events: there are no absolute subjects or objects. At the conclusion of these studies, I submitted an honors thesis that was later published in part as the book *Choosing Reality: A Buddhist View of Physics and the Mind.*

As satisfying as I found these studies at Amherst, I felt a growing urge to engage wholeheartedly in meditation once again. So, shortly after graduation, I went into retreat in the high desert just east of the Sierra Nevada, where I had backpacked as a youth. Living alone in the wilderness, I now faced one of the toughest choices in my life. I had lived as a monk in the West for more than five years, with no monastery, abbot, or fellow monks. To a large extent I felt alienated from my native civilization, which provided little support to those devoting themselves to a monastic way of life. What direction was my life to take now, I wondered. After much soul-searching, I decided that I would continue to live in modern society, seeking to help restore harmony and balance to a world that appeared even more fragmented than when I had left it sixteen years earlier. Since this appeared to be largely incompatible with the monastic way of life, I formally returned my vows in the spring of 1987, bringing to an end my fourteen years as a Buddhist monk.

As one way of life closed, another opened. In the autumn of that year, I was invited to serve as interpreter in the first Mind and Life Conference when the Dalai Lama and a select group of scientists met to discuss the nature of the mind in the natural world. The proceedings from this first meeting in Dharamsala were later published in the work *Gentle Bridges: Conversations with the Dalai Lama on the Sciences of the Mind.* I have subsequently served as co-interpreter with Thupten Jinpa in virtually all the dozen Mind and Life conferences since then. Over the years, topics of these meetings have ranged from the mind-body problem and consciousness and emotions, to cosmology, modern physics, and the nature of matter and life. In each of these conferences scientists at the cutting edge of research have shared their insights with the Dalai Lama and other Buddhist contemplative scholars, and they, in turn, have learned about Buddhist ways of experientially and rationally investigating the nature of reality.

Such encounters inspired a return to academia. While I was formally enrolled in Stanford's doctoral program in religious studies, research took me into the fields of cognitive psychology and the philosophy of physics, biology, and mind as well as comparative religion. The historical interface between science and religion in the West was especially intriguing as it pertained to the scientific study of the mind. This research resulted in a book *The Taboo of Subjectivity: Toward a New Science of Consciousness,* and a doctoral dissertation on the contemplative training of attention was eventually published under the title *Balancing the Mind: A Tibetan Buddhist Approach to Refining Attention.*

Shortly after matriculating at Stanford, I had the great good fortune of becoming a student of Gyatrul Rinpoche and for the next ten years served as his principal interpreter. He gave me extensive instruction in the Nyingma and Kagyu orders of Tibetan Buddhism, with a special emphasis on the "Great Perfection" (Dzogchen) tradition. Through this close relation to him, I was able to meet many other masters from this lineage, including Dodrupchen Rinpoche, Khenpo Jigme Phuntsog, and Yangthang Rinpoche. The essence of the Great Perfection is the natural liberation of the mind through realizing in a nondual way the fundamental nature of awareness. This, I believe, is the premier science of consciousness.

After teaching at the University of California at Santa Barbara for four years, in the fall of 2001, I returned to the high desert east of the Sierra Nevada and spent six months in solitary practice, which turned out to be a time of unprecedented integration. All that I had learned during my first twenty years growing up in America and Europe as well as my religious, philosophical, and scientific training over the next thirty years were woven together into a tapestry of ancient and modern, Eastern and Western strands. I could now see a way to thoroughly integrate the three themes that constituted a meaningful life: the pursuit of genuine happiness, understanding, and virtue. My worldview, meditative practice, and way of life now coalesced into a unified whole, and for the first time I felt no dissonance among the many influences that had formed me as an individual.

As I prepared to emerge from this period of solitude, I reflected on how I could best be of service to the world, given my diverse background and the many wonderful scientists, contemplatives, and scholars who had shared their knowledge and wisdom with me. This led me to envision the Santa Barbara Institute for Consciousness Studies. One of its main projects is to integrate first-person inquiry, found in the world's contemplative traditions, with the third-person methods of modern science. Until now, cognitive scientists have sought to understand consciousness by studying the mind in terms of its neural bases and expressions in behavior. Because of the limitations of this approach, which examines states of consciousness only through their physical correlates, researchers are predisposed to view the mind from a materialistic perspective. The time has now come to apply rigorous methods that directly examine the broadest possible range of mental phenomena, including those experienced by highly advanced contemplatives. I am convinced that such an approach, combining the rigor of both scientific and contemplative inquiry, can bring about a true revolution in the mind sciences. This revolution may not only yield profound revelations concerning the nature and origins of consciousness but may also shed new light on the possibility of exceptional mental health and well-being and, ultimately, spiritual awakening. I believe that humanity is now on the verge of a new renaissance, bearing a unique opportunity to discover the hidden harmony of science and spirituality.

B. Alan Wallace

Acknowledgments

THE WRITING of this book was itself an interdependent venture, and we would like to acknowledge the collaboration of Carl Chew, Dr. William L. Ames, and Daniel Hodel, who all offered valuable suggestions after reviewing an early draft of this work, along with the distinguished physicist Dr. Fred Cooper, who reviewed a later draft. Special thanks also goes to Shambhala editor Michele Martin, whose ideas and enthusiastic support helped craft this book into its final shape.

Introduction

How do we know things? How do we decide that something is true? Since its appearance four hundred years ago, science has tried to answer this question by focusing on the physical elements of the universe. Many scientists now believe that physical phenomena alone are real and that we know this is true through objective scientific investigation. We are told that (1) the universe is exclusively physical, (2) this is a proven fact, and (3) we learn all of the important things about reality by virtue of science—period.

Actually, what was just described is a myth. It was never arrived at scientifically. Rather, it resulted from a process that automatically filtered out contradictory evidence. In certain areas of science, physics in particular, this exclusively physical view of reality was brought into question a century ago. Even so, the consequences of considering nonmaterial phenomena as real are upsetting to many scientists. Science, after all, distinguished itself from religion by denying nonphysical explanations of phenomena, such as miracles and demons. Therefore, even as scientific theories based on nonmaterial hypotheses find application in technology we use daily, their implications for our understanding of reality are generally ignored. For example, photocells and computer chips were developed from scientific theories that don't jibe with a purely physical view of reality. We use and accept them, but we don't often explore their scientific basis and its implications.

Part 1 of this volume—"What's Wrong with This Picture?"—is a detailed deconstruction of this materialistic myth. Science was born in the largely Christian world of Renaissance Europe; therefore, it is not surprising that even as science negated many Christian beliefs, it was unavoidably wedded to basic tenets of Christian theology. Over time, as it gained strength and credibility, science itself took on aspects of a religious dogma. Burdened with such entrenched beliefs yet beholden to its experimental method—which sometimes produced evidence that contradicted its materialistic outlook—science developed a split personality. Since the turn of the twentieth century, when the new physics upset the apple cart, this schizophrenia has hobbled science in its quest for a thorough understanding of the nature and origins of the universe.

The source of this problem from the very beginning was the scientific attitude toward the mind. Since mental phenomena, such as consciousness, thoughts, images, and emotions, were not physical, the mind had to be either ignored or regarded as a property of matter. Because nonmaterial phenomena were unacceptable, they couldn't be studied scientifically. By the same token, the mind could not "exist" unless it was reduced to something purely physical, the "gray matter" of the brain. That was the obvious solution. Nevertheless, by the early twentieth century it was becoming clear to physicists that certain properties of matter itself depend on the role of the observer—mind and matter are not independent from each other. Furthermore, present-day attempts by neuroscience to pin down the mind as purely physical are riddled with problems. Could it be that both mind and matter are real, part of nature, and intimately interconnected?

In part 2, "Consciousness: Completing the Picture," we discover that contemplative spiritual traditions have long been studying this problem and have come up with some refreshing hypotheses. By accepting the validity of nonmaterial phenomena—the mind especially—they suggest we may arrive at a complete and harmonious view of reality. Science has made a deep exploration into the material nature of phenomena. Contemplatives have studied the mind that observes phenomena and that does the investigating. Conclusions drawn from both of these traditions point to a middle ground, a universe of *interdependence* between mind and matter. That suggests science and spirituality might work together in a complementary fashion to arrive at a more embracing view.

Yet do scientists, with their rigorous methodologies, have enough in common with contemplatives for the two to collaborate successfully? We shall see that in fact a number of spiritual traditions, particularly those from Asia, do have rigorous standards of objectivity for exploring the mind. Although this is not widely known in the West, meditation has been developed over millennia into a precision instrument of observation. The aims of this contemplative science may be spiritual—release from psychological suffering, attaining enlightenment, inner peace, and so forth—but the means of accomplishing those ends require a clear understanding of all phenomena. Here a human being is not seen as a stranger in an alien universe but as a full participant, intimately interconnected with everything. His or her achievement of inner freedom therefore requires a deep understanding of the whole. This picture is further elaborated in part 3 with a description of one such science of consciousness, from Indo-Tibetan Buddhism. Here is a view that may illuminate even some of the most exciting current theories from physics about the origins of the universe. Where the informed scientific imagination can merely describe such possibilities, meditators claim they directly experience the creative forces at play in an ultimate reality accessible to highly refined states of consciousness.

There is, and always has been, a single underlying basis common to both science and spirituality. This common ground—that which makes us human—is the mind. It is to human minds that myriad phenomena appear, and it is the human mind that has explored nature, beheld scientific discoveries, and formulated theories to account for them. But just as spirituality often goes too far, imbuing the mind, spirit, or soul with a mystery that clouds our understanding, so has science also gone too far in stripping the mind down to the bare chassis of the brain. This volume was written to help balance our understanding of the mind, showing that it is indeed wondrous, but that its extraordinary qualities can be understood without sacrificing our intelligence.

PART ONE
What's Wrong with This Picture?

Y OU'VE PROBABLY seen those cartoons of animals in a jungle or crowds at a carnival—with the caption below it reading, "What's wrong with this picture?" At first glance the scene seems perfectly normal. Yet when you inspect it carefully, you begin to discover oddities. There's a man with two heads, an elephant with three trunks, or a giraffe wearing shoes.

The picture of science presented in the classroom and in the media—what it's supposed to be and how it developed—also seems plausible at first glance. In part 1 of this book we are going to discover some very strange creatures hidden in the foliage— oddly unscientific ideas that blend in with and even guide much of scientific thinking. Just as important, items that we would expect to see have somehow been left out.

A Shotgun Wedding
The Marriage of Science and Christianity

1

"*SEEING IS BELIEVING!*" We hear all the time. In our normal view of things what could be more obvious? There seems to be no question that the objects that enter our field of vision are really *there*. That's why we tell our children to watch out for traffic when they cross the street and why we duck when someone throws a snowball our way. Visual confirmation is crucial for a wide range of day-to-day activities: In business we keep our eyes, literally, on the bottom line. As proof that something is true, we want to "see it in black and white." When we buy a home or a car, although we may be attracted by a sales pitch, we would never write the check without having carefully looked over the merchandise.

Proof that things are real extends to the other senses as well. We believe in things that we can touch, especially when we burn a finger while cooking or bang our head on a low-hanging branch. What could be more real than the smell of fresh-cut grass, the voice of a loved one on the telephone, or the taste of our favorite dessert? We have no doubt that these things exist, so generally speaking *sensing* is believing.

A SIXTH SENSE

Beyond our five senses we also perceive things "in our heads," with our mind's eye. We rely constantly on fleeting mental images that appear to us like some personal, internalized video monitor. Furthermore, the impressions we experience go well beyond the visual. We imagine that we see, touch, hear, smell, taste, and think—even in our dreams. Here we don't actually perceive objects with our sense organs, but we do possess a mental perception that gives us that impression. These internally perceived objects have a strong sense of reality. When dreaming, we are convinced that these mental phenomena are real. Rarely if ever do we stop in the middle of a nightmare and say, "I don't believe this is really happening—I must be dreaming."

In the daytime we use this theater of the mind, this interior space, to organize our lives: to imagine how to get from downtown to the airport, or to figure out how much change we should receive from the cashier. The whole inner, or subjective part of our lives is played out in this arena. Hopes and dreams start and sometimes end there: "Does he love me? Does she not?" Since it makes us aware of this inner world of thoughts, feelings, mental images, dreams, and so forth, *mental perception* qualifies as our sixth sense.

Upon reflection it is clear that mental perception is a bit different from the other senses. The interior objects it perceives have no concrete, physical reality. The seeming reality of a dream vanishes when we awaken, and our daydreams don't have the immediacy of a burned finger or the coins handed to us at the checkout counter. But mental perception and imagination play a vital role in a second way we know things: by reasoning, or inference.

We wouldn't get very far if we relied exclusively on our senses to know things, would we? Lounging in a hammock, half asleep, you hear a high-pitched squeal and then a crash. It must have been an auto accident. You didn't see it, but you know the signs and can put two and two together. Our imagination provides a mental drawing board where we can make calculated guesses about the reality of things, reason things out. The objects we manipulate—images, thoughts, emotions, and so on—are illuminated by our faculty of mental perception. By thinking, remembering, and inferring, we use such symbols drawn from experience to enhance our knowledge and, hopefully, to make our lives better.

Science is a good example of this process. First a problem is presented —something mysterious or not clearly understood. Next a theory regarding the problem is devised in the scientist's mind, based upon earlier theories and experiments and the free flow of the scientific imagination. In the beginning this may be no more than an informed guess. Later the theory may be altered to fit new data gathered experimentally with instruments such as the microscope or telescope, which extend the power of our senses. That, for example, is how the planet Neptune was discovered. Prior to its visual confirmation with a telescope, its presence was theorized on the basis of deviations detected in the movements of a neighboring planet. Shortly afterward, using the theory as a guide, Neptune's existence as a planet in our solar system was confirmed visually.

This brings us to a third way we know things—knowledge based on authority. No human has yet seen Neptune with the naked eye, and very few of us have seen it with a telescope. Yet we accept the authority of scientists who tell us Neptune is there. When shown a photo of Neptune, few of us are likely to say, "I don't believe Neptune is real. These astronomers may be showing us a doctored picture of a Ping-Pong ball." In the same way, we believe that a man named Napoleon Bonaparte once ruled most of Europe and that viruses can cause a cold. We've never met Napoleon, and few of us have studied microorganisms in a laboratory, but we accept the word of astronomers, historians, doctors, and many other authorities because we believe they have special expertise that we lack. According to conventional wisdom, they are reliable.

We could sum up the three main ways we know things with the following scenario: Glancing at the street outside my window, I see a pedestrian fall to the ground in agony after being sideswiped by a car. I call 9-1-1 and tell the operator what I saw. The pedestrian experienced the pain of the accident directly with his *senses*, mainly the sense of touch. With my eyes I witnessed the expression of pain on the face of the pedestrian and saw him fall to the ground gripping his leg. I may have also heard the pedestrian scream. However, the fact that the pedestrian was actually injured by the car, that he was in pain, was only an *inference* on my part. He might have been faking it as a prank, or he and the driver may have been in collusion to defraud the insurance company. The emergency operator believes me on *authority*, assuming I am an honest citizen with my senses intact. A few pertinent questions will convince the operator that I am a credible witness and not a prank caller.

The acquisition of knowledge can also be viewed in terms of the degree to which objects are hidden from us and our capacity to understand them. For the man experiencing the accident, the object in question—his pain—was obvious, it wasn't hidden from him in the slightest. For me, the witness, his pain was slightly hidden in that I had to infer it from experience, for instance from knowing my own facial expressions when I suffer an injury or what I might feel like if I fell from a bicycle. The pedestrian's pain was very hidden for the 9-1-1 operator. He had neither direct experience of it nor enough evidence to make an inference. He had to take my word for it.

HIDDEN TRUTHS

This last category—very hidden knowledge—covers many of the most mysterious and important aspects of life, answers to big questions such as the origins of the universe, the existence of a creator and of fate, life after death, and so forth. We all believe (or disbelieve) in things that are hard to prove on the basis of the senses or reasoning. Such beliefs may come from our family or community—we have learned to believe what they believe. Surrounded by others who share our beliefs, we take these same people as authorities. Beliefs may also come from the truth and wisdom we perceive in teachings of some prophet or religious leader. At some point in our lives the words of Muhammad, the Virgin Mary, Jesus, Lao-Tzu, Mahatma Gandhi, Buddha, or some other spiritual authority may strike a chord within us, inspiring us, connecting with our own thoughts, experiences, and views about life. "Ah-ha!" we say. "This person knows the truth."

We may accept that person as the ultimate authority on reality, perhaps because we believe him or her to be "divinely inspired," to be indivisible from the "one true God," or to be endowed with "perfect wisdom." A Christian may believe in the existence of heaven because Jesus said it exists. A Hindu or Buddhist may believe in reincarnation because authorities such as the Hindu saints or the Buddha said it is true. One could speculate that prophets and saints are able to penetrate these mysteries directly with the senses or by some form of reasoning. But such spiritual truths are hidden from most of us. With rare exceptions only the most holy or gifted are said to perceive them.

As far as the general public is concerned, much scientific knowledge also falls into this class of hidden information. As we saw in the case of the injured pedestrian, the degree to which knowledge is hidden depends upon the perceiver. Prior to the seventeenth century, Neptune's existence was a very hidden truth to all but Galileo, who in 1612, using a telescope, saw a vague something in the night sky that we now know to be Neptune. In 1843, when the British astronomer John Couch Adams deduced Neptune's existence from mathematical calculations of the orbit of neighboring Uranus, the truth of Neptune's existence became somewhat less hidden—it became a scientific theory, believed by some astronomers to be true by inference. After 1846, the year Neptune was correctly identified with a telescope, its existence as a planet within our solar system became an accepted scientific fact for all but the extreme skeptic.

Accepting the existence of Neptune was comparatively easy. Many of the most important and fundamental goals of scientific understanding are as hidden from us even today as proof of the existence of heaven or reincarnation. Recall the story of Isaac Newton and the apple. Prior to Newton's theory of gravitation, when an apple separated from its branch and moved toward the earth, it was said that the apple "fell." Anyone could tell directly with the senses that the apple was heavier than a feather or a dust mote and would therefore fall rather than be blown by the wind. Apples, stones, and cannonballs had weight, so they fell. Few people bothered to ask what weight was. Newton did, and his conclusion was that weight resulted from a force of attraction. The apple didn't fall, it was attracted to the earth. But why?

According to Newton, the mass of any given object emanates an invisible force called gravity, creating a gravitational field that attracts other objects to it. Newton formulated his gravitational theory mathematically. Gravitational fields cannot be comprehended directly with the senses, but rather only through inference. Once Newton's theory became a scientific law, having been confirmed by experiments and approved by the scientific community, the general public accepted it on authority. For those unable to confirm it through experiments or by understanding its mathematics, gravity was a very hidden truth accepted on faith in the scientific community, on authority.

This question of how we know what we know and believe what we believe becomes downright puzzling when we turn to a more recent theory of gravitation. In the early twentieth century Albert Einstein theorized that gravity is a feature of the curvature of space-time. The apple isn't attracted to the earth, but rather follows a path of least resistance built into the shape of space itself. Massive objects like the earth happen to create a particular space-time curvature that leads objects such as "falling" apples "downward." For physicists the gravity of curved space is a hidden object of knowledge understood through inference. Some members of the general public familiar with the theory of curved space may accept that explanation on the authority of Einstein, the most famous genius of the twentieth century. He had an excellent track record, they reason, so his theory, however strange it might seem, should be believed, for it has been corroborated by many other scientists as well. Some of us may hold to the classic Newtonian view, which has had several centuries to become widely known. But when we slip in the shower, Newton and Einstein are nowhere to be seen. We *fall*. Likewise when we watch Olympic weight-lifters grimacing and straining on barbells, or when we ourselves try to lift a heavy object, we all, even physicists, sense the *heaviness* of these objects rather than a gravitational field. Such is the force of habit.

So for falling apples, which is it—Newton, Einstein, or the habitual beliefs we call conventional wisdom or common sense? If scientific theories didn't lead to new capabilities and inventions, such as landing human beings on the moon, theories of gravity might sound like fairy tales or mere speculation. No one has yet seen gravity, so it remains, like heaven and reincarnation, a hidden object of knowledge. Just as with heaven and reincarnation, there is also the possibility that gravity, as such, is either unknowable to us or does not exist at all.

SCIENCE SAYS TO RELIGION, "MAY I HAVE THIS DANCE?"

For Western civilization back around the time of the Renaissance, hidden knowledge was obtained from many sources. Astrologers claimed to know the future from the stars. Tradition dictated the best time to plant crops. However, the main place to look for hidden knowledge about the big metaphysical questions was the Bible. The nature and origins of the universe and human beings' part in it were all explained in this collection of stories written or handed down from Hebrew prophets, Jesus, and apostles who claimed inspiration from God. For the average Christian, God's word was the ultimate authority. So powerful and widespread was this belief that those who questioned it could be brought before the Inquisition, where excommunication, imprisonment, torture, or execution awaited.

Nonetheless, from the fifteenth through the mid-seventeenth centuries fresh ideas from the Arab world and from the rediscovery of Greek learning gradually transformed all aspects of European culture, including religion. Philosophy and the arts flew off in new, more secular directions. Education, too, gradually became secularized as universities replaced monasteries as the centers of learning. Commerce succeeded feudalism, new continents were discovered, city-states and nations gained power, while the theocracies and monarchies of the Middle Ages lost influence. Even ideas that challenged the Bible were cautiously advanced. The most important child of this European rebirth was science.

Usually we think of science as an orderly process of investigation through which theories must be proven experimentally. Inspiration plays a part, but as Thomas Edison said, genius is 99 percent perspiration—the hard work of devising new instruments, running experiments, and doing the math. This modern notion of science is summed up in the scientific method, which can be defined as "principles and procedures for the systematic pursuit of knowledge involving the recognition and formulation of a problem, the collection of data through observation and experiment, and the formulation and testing of hypotheses."[1]

Ideally, once a theory has been proposed and convincing evidence gathered, the originating scientist or scientific team publishes its findings in a scientific journal. This initiates the process of peer review, where other scientists carefully examine the original work, debate it, and try to reproduce its results. Once this gamut of tests has been successfully run, the new theory is accepted, though only tentatively since new theories appear frequently to challenge the old. The ideal for the process of scientific investigation was given by the physicist Richard Feynman: "Experimenters search most diligently, and with the greatest effort, in exactly those places where it seems most likely that we can prove our theories wrong. In other words we are trying to prove ourselves wrong as quickly as possible, because only in that way can we find progress."[2]

However, the science of the Renaissance did not emerge from the womb full-grown, with a guidebook to the scientific method in hand. Today, science is generally viewed as the antithesis of religion—facts proven experimentally as opposed to beliefs accepted on authority. Yet the development in the Renaissance of such a science, one completely divorced from religion, is inconceivable. In actual fact, science was the product of a highly religious society. The pioneers of science—men such as Copernicus, Descartes, Galileo, and Newton—were Christian believers. Many of the fundamental ideas of science are found in the writings of Saint Thomas Aquinas (1225–74), who assimilated the ancient Greek philosophy of Aristotle into Christian thought. Therefore, it was inevitable that science would develop initially along Christian lines. Then as science evolved and its theories and experiments produced proven, practical results, a dialogue, often political in nature, took place between scientists and ecclesiastical authorities—a cautious dance of give and take. Much was at stake.

Through the course of this exchange, the church was gradually forced to modify its opposition to science, so much so that by the twentieth century science had for many people replaced religion as the final authority on reality. Even so, scientific thinking never completely divested itself of ideas derived from Christian theology. They were too deeply embedded. As a consequence, the prevailing *popular* view of science in the West is based on the discoveries achieved by the scientific method, but infused with a hidden Christian view of nature. That view evolved from a set of metaphysical assumptions that underlie science and are believed by many scientists today—they're collectively called *scientific materialism*.

The philosophy of scientific materialism, which emerged clearly only in the nineteenth century, is an interpretation of science. It is based on five principles: Stated succinctly, the first and foremost is *objectivism*, which states that the only reality of importance exists "out there" beyond our minds—the objects we perceive as the physical universe. Next is *metaphysical realism,* the belief that the objective universe can be known by the subjective human mind. Objectivism and metaphysical realism were further refined by the *closure* principle, which denies the possibility that anything other than material influences can affect any aspect of the natural world. Fourth, the principle of *universalism* declares that these rules are universal—they are the same in every corner of the universe be it the center of a cell or the center of a star. Finally, *physical reductionism* reduces all of nature to physical entities and interactions. Combining these five principles, this materialistic view of reality was the unintended offspring of the marriage of Christianity and science.

In the Beginning

With its Christian background it is hardly surprising that science was strongly influenced by the Bible, which opens in Genesis with the words: "In the beginning God created the heavens and the earth." On succeeding days God elaborated the physical universe, then populated it with the animals, and not until the sixth day, the last day he worked on his Creation, did he make humankind. And God made humans "in his own image." Later, when Adam ate from the tree of knowledge, that image was tarnished: the fall of humans from God's grace. This is the official starting place for the beliefs of Judaism and Christianity, and science too can be traced to this source.

Given these fundamental beliefs about the nature and origins of the universe, it was perfectly natural for the founders of science to see their main task as using the mind of humans (made in the image of God) to fathom the real, objective world created "out there" in the heavens by him. More specifically, since God also "governed" the universe, scientists sought to understand his "laws" of nature. This set the foundation for *objectivism,* which declares that there is an independent, objective reality outside of our minds, beyond our thoughts, and that this is what science aims to understand. The goal is to fathom the material stuff called the universe. Whether that is to be found in the chemistry of cells, the forces of subatomic particles, or in distant galaxies, all of these phenomena are "out there" beyond our inner, subjective thoughts and feelings, just as the heavens were to Renaissance astronomers. Objectivism is the main principle of scientific materialism.

That seems perfectly natural, doesn't it? Whether the field is physics, chemistry, or biology, research is performed on material objects such as atoms, chemical compounds, or cells in order to understand how the universe ticks. Our image of scientists in their white coats, hovering over their instruments, making their calculations, includes the notion that they are seeking diligently to solve the riddles of physical existence. However, this view, this pattern of thought called objectivism, is really no more "natural" than reading tea leaves to tell the future. It has its roots in the Bible and seems obvious only because it is such an ingrained belief. If instead the universe were believed to be a fully integrated whole, as it was in ancient China, then reading tea leaves (or more likely reading patterns of sticks or coins, as was done in the *I Ching* system of divination), would itself seem perfectly natural. Just from observing the world's many cultures and beliefs we know that the naturalness of things depends largely on one's beliefs, reinforced by tradition—the force of habit.

But what if the tea leaves say, "Don't believe the tea leaves"? What are the consequences of seeking the objective truth of "God's domain" if the experimental results obtained differ from the beliefs held by religious authorities—"God's representatives"? As it turned out, seeking an understanding closer to God's led early scientists to conclusions that contradicted the cherished (and officially recognized) views of popes and cardinals. Those beliefs held sway, and prelates could use extreme measures to protect them. The conclusions of scientific experiments designed to understand the mind of God could be interpreted as heresy.

Since the original goal of science was to understand the universe "out there," astronomy, heavily influenced by biblical cosmology, became the first science of importance. Early astronomers studied the heavens and began to investigate the motions of the stars and planets, hoping in that way to understand God's Creation. Astronomers set out to achieve God's perspective on the universe by using an "impersonal" technology—something independent of the sinful human body—the telescope.

Prior to the Renaissance, the earth-centered model of Ptolemy, which was in perfect accord with the Bible, dominated astronomy. The basic idea was that all the planets moved in circular orbits around the earth. A complicated and ingenious explanation using epicycles (circles within circles) accounted more or less for observed planetary orbits. (Viewed from the earth, which we know to be just another planet circling the sun and not the center of the solar system, the observed motion of other planets is complex and squiggly—not simple circular motion.) However, as time passed, trying to decipher planetary motions with Ptolemy's model became a frustrating affair. When new observations deviated from the theory, new, ever more complicated epicycles had to be devised. It just wasn't working.

Seeking a more accurate method, Copernicus, in the early sixteenth century, introduced the heliocentric system: it was the sun, not the earth, that remained at rest as the universe moved about it. This model accounted more simply for the motion of the planets, their order in the solar system, and their distances from the sun. In essence it also introduced a "God's-eye view" of the universe to science, though it might better be called a "sun's-eye view." If you were on the sun, Copernicus reasoned, this is the relative motion you would see: the sun, like God, is at absolute rest, while we on earth spin around it in circles. God, after all, did not view the universe from earth but from the heavens themselves, from a more objective standpoint. Suddenly, human beings' self-centered notion of their own importance—a subjective influence—would be replaced by God-like objectivity. The Copernican view therefore amplified the ideal of objectivism. Not only was the scientist to study exclusively the objects of the universe, but subjective distortions due to human beings' imperfections and limitations were to be eliminated.

There was a problem, however. The heliocentric theory could be interpreted as placing earth—home to sin, war, and pestilence—in the heavens. This was in conflict with a literal interpretation of the Bible. Copernicus, whose theory wasn't published until he died in 1543, wasn't around to get into trouble with the church over his potentially heretical theory. Certainly, as a church official (with a doctorate in canon law), he was aware of the theory's conflicts with theology. This was one reason he delayed its publication for thirty-six years. Galileo, on the other hand, who built the first astronomical telescope and vigorously advocated the heliocentric model, got into very hot water with the authorities.

In 1616 the church declared Copernicus's theory heresy. Subsequently Galileo, as its most avid and articulate proponent, was brought before the Inquisition in Rome and forced to renounce his belief in the Copernican model. He publicly accepted an explanation put forth by a church official, Cardinal Bellarmine, that the heliocentric model merely made astronomical calculations easier but did not represent reality. The Copernican model was officially reduced to the status of a useful gadget for calculating planetary geometry, that's all. This was a severe personal blow for Galileo. It was several years before he had the courage to reenter public debates on astronomical matters. Galileo's story illustrates the power of the Catholic Church at that time to modify scientific conclusions, at least publicly. Galileo's most important writings on astronomy were kept on the Index of Prohibited Books until 1835 (as were those of Copernicus and of Kepler, who improved the heliocentric model). Full acceptance of objectivism would have to wait.

By the Numbers

Galileo was the greatest scientific genius of his time. He believed, as did Plato and Augustine before him, that God ordered the universe through numbers, that mathematics was the divine language. He put it this way:

> Philosophy is written in this grand book, the universe, which stands continually open to our gaze. But the book cannot be understood unless one first learns to comprehend the language and read the letters in which it is composed. It is written in the language of mathematics, and its characters are triangles, circles, and other geometric figures without which it is humanly impossible to understand a single word of it.[3]

Again, since humans were made in the image of God, scientists, by deciphering the universe in God's own language, mathematics, might also reveal God's way of thinking the universe into existence. This belief in a mathematical foundation of the universe became a cornerstone of science. It is an essential component of *metaphysical realism*, another of the principles of scientific materialism, which states that the universe is ordered by ideas (such as mathematics) that lie beyond or transcend the senses. According to this view, these ideas are implicit in the very nature of reality, what the universe really boils down to. We know that different species perceive similar objects differently, and there are variations in how different people perceive the same things, such as colors and sounds. But what are these objects like really—objectively, independent of the perspective of any of God's creatures?

As with the workings of gravity, the transcendent reality of phenomena must be either inferred through a process of reasoning or accepted on authority. A scientific theory is a transcendental idea that aims to account for reality as it exists independently of our human experience and concepts. Again, the origin of this way of thinking can be found in scripture, the "grand book" that shaped the minds of Europeans prior to and during the Renaissance. So, just as God transcends the universe, his divine language (and later, scientific theories heavily dependent on mathematics) allows the scientific mind to transcend the senses and reach true understanding of reality. Often, according to this perspective, reality can be reduced to a mathematical formula.

Metaphysical realism seems just as natural as objectivism. We've all had a taste of geometry and its laws. Using pencil and paper or calculating in our heads, we can figure out the square footage for tiles we wish to lay on a bathroom floor. We don't have to rely on our senses and measure tile by tile. We can also calculate how many miles to the gallon we got on our last automobile trip, or the rate of growth of our business. Here the knowledge we attain about things is completely abstract. No one has ever seen or touched a "mile per gallon" nor a "one percent yearly increase in profits." Yet they are a perfectly natural part of our everyday reality.

Let's Be Practical

Of course science in the Renaissance was not propelled merely by the curiosity of talented intellects. Expanding commercial interests demanded improved technology in the field of navigation, and governments needed better military engineering to protect commerce and advance the interests of their domains. As science provided this technology, it gradually gained enough credibility to prevent the church from censuring it each time a new theory or discovery contradicted religious doctrine. Rather, a delicate balance replaced the previous religious dominance over scientific ideas. Once the prestige of science had grown, theologians themselves began to rely on it. For example, science could be used to determine what was and was not a miracle: if some phenomenon could not be accounted for by a scientific law, it could be declared miraculous. Over time, as scientific knowledge expanded and God's role diminished, he would be called on merely to fill in the gaps in scientific knowledge—gaps that grew smaller and smaller as science advanced.

At this point, in order to provide technology as well as to effect a more perfect understanding of the heavens, specific knowledge of the laws of motion was required. The artillery captain needed to calculate the proper angle for his cannon in order to destroy enemy positions. Presuming that the motions of cannonballs and planets followed the same laws, the discovery of those regularities was a major interest for such leading scientists as Descartes, Leibniz, and Newton. One important step in that direction was to determine whether nonphysical factors played any part in the motion of objects. René Descartes (1596–1650) was the first to theorize the *closure principle* when he stated that there existed nothing in nature that could not be explained by "purely corporeal causes." That idea, another principle of scientific materialism, closed off nature from all but physical influences. Frightened soldiers couldn't cause cannonballs to fall elsewhere by wishing or praying. Nor could demons start fires or cause objects to levitate. Only matter could move matter. However, Descartes was careful to make two exceptions: biblical miracles and the human soul, which he believed could affect the body. But the influence of the church had already weakened considerably. Leibniz, born only four years after the death of Galileo (1642), boldly theorized that mind and spirit had no effect whatsoever on nature.

SENDING GOD INTO RETIREMENT

Isaac Newton believed the physical universe was composed entirely of inert matter created and put into motion by God, who imposed his laws upon it. Deeply religious, he argued that Leibniz's view, a universe completely self-contained and isolated from spiritual influences, would lead to materialism and atheism. But it was Newton himself who laid the ground for a completely mechanical model of the universe, one that God may have set in motion but that no longer required him to keep running. Scientists who succeeded Newton added to this mechanical model the belief in *universalism*, the principle that natural laws were the same everywhere and at all times, and *physical reductionism*, the idea that nature could be reduced to physical entities and their functions, each totally isolated from the rest, having no connections save the patterns imposed by the laws of nature (whose author was God). Everything could now be reduced to matter and its laws and properties.

In his *Principia* (1687), a treatise considered to be the foundation of classical physics, Newton set down his three famous laws of motion. They deal with the movement of objects in terms of inertia (the tendency for objects to remain either at rest or in motion until acted upon by an outside force), changes in velocity, and the effects of contact between two bodies (yielding equal and opposite reactions). The laws of motion led to Newton's theory of universal gravitation, and that put science on a roll for over two centuries. A whole series of scientific laws based on Newton's discoveries became known as classical physics. With these in hand one could build useful devices such as airplanes and guided missiles, or improve one's golf game.

Today, whether we have studied them or not, the laws of classical physics seem intuitive. Their best-known simplification is the analogy to billiard balls: ball A hits ball B at angle x and velocity y, causing ball B to move in direction d at velocity z—all very mechanical. In spite of the fact that physics has now gone far beyond this simplistic model, this style of cause-and-effect thinking still permeates the modern mind as a component of common sense. Even atoms are generally pictured as little balls in various arrangements resembling planetary systems. Whether it be atoms, the structure and functioning of machinery, the dynamics of weather systems, or the logic boards of computers, in the public's imagination "reality" is mechanical in nature. To understand reality, dissect it into its various parts and study how the parts interact in terms of cause and effect: billiard ball A hits ball B . . . —and the truth is revealed.

As one might expect, the many laws of classical physics, built upon the foundation laid by Newton and his contemporaries, mesh nicely with the five principles of scientific materialism. Even though Newton and his early followers had never heard of "scientific materialism," they were guided by some of its tenets. They sought an objective view of an objective universe. They discovered laws, defined mathematically, that appeared to describe the essence of the material universe (metaphysical realism). Although Newton was in some degree a believer in the miracles of Christianity, later scientists gradually abandoned these beliefs, embracing the closure principle theorized by Descartes. The laws of classical science were assumed to be universal, and the universe could be reduced to physical entities and their interactions (universalism and physical reductionism).

THE WALL FLOWER

But as science and religion twirled and curtsied in their dance for supremacy, someone else in attendance was not invited to dance. What about the human mind, the originator of scientific theories, the primary scientific instrument? Although created in the image of God, it wasn't part of the objective, physical universe. Why was an entity so prominent as the mind treated from the beginning like a wall flower, something that's obviously involved, yet excluded from the domain of scientific inquiry?

Here's why: the mind may have been capable of aspiring to understand God's kingdom, but many of its thoughts and emotions were impious. This was, after all, the mind of humans *after the fall of Adam.* The nature of human beings' mind or soul was that of original sin. And especially since Renaissance Europe had just emerged from two centuries of witch-hunting, the dark inner sanctum of the mind was highly suspect. It was this subjective aspect of human beings, especially their imagination, that harbored the evils inspired by the Devil. Such views are confirmed in the writings of popular Renaissance philosophers. Francis Bacon, an important philosopher of science, claimed that science was sanctioned by scripture as the means through which humans could reclaim their dominance over nature, lost through the fall of Adam. So as long as the mind was "scientific," facing outward, exploring the objective universe created by God, all was in harmony with Christian beliefs. But the inner, slithery, subjective realm of the mind was neither part of nature nor to be trusted or ultimately believed in. As far as science was concerned, the mind got no respect. As we shall see later on, this is in stark contrast to views developed in Asia, where the mind has often been the principal topic of interest.

The European tendency to mistrust the subjective, imaginary, mental realm only increased with the Protestant Reformation at the beginning of the sixteenth century. The new religious reformers condemned the priestly magic of the Catholic Church—the sacraments and saintly miracles—and warned of the dangers of diabolical influences on the mind. Moreover the Protestant ethic, aimed at humans improving their lot in the world through hard work, mistrusted magic, not only because it was an inner phenomenon but also because it was a kind of easy way, a shortcut to achieving one's ends. If a magic spell could bring one wealth, why work? It is not surprising, then, that three centuries would have to pass before an experimental science of the mind, psychology, would emerge in the West.

This sword of mistrust cut both ways. The gradual decline in the belief in magic was accompanied by a questioning of God's role as a miracle-worker. It was a painful dilemma: If God could intervene at will by magically producing miracles, a universe of consistent natural laws based on the closure principle, universalism, and physical reductionism was illogical. On the other hand, if this mechanical model of the universe didn't need God, it was heresy.

Even so, science was soon to squeeze everything nonmaterial out of the universe —spirits and demons, the human mind, and God himself. Before long, the astronomer Pierre Laplace (1749–1827), when asked about God's role in the world of nature would say, "I have no need for that hypothesis."

THE TRIUMPH OF MATERIALISM

By the beginning of the nineteenth century, with the road now relatively clear of philosophical and religious obstructions, classical physics provided the basis for a string of important discoveries that inspired scientists to believe they could account for the entire universe using sophisticated mathematics.

In the physical sciences Joule, Mayer, and Helmholtz devised the principle of the conservation of energy. This evolved into the first of the three laws of thermodynamics, revealing the relationships between heat and energy and giving them mathematical formulations. The work of Augustin Jean-Fresnel led to the wave theory of light. Lord Kelvin and James Clerk Maxwell gave magnetism and electricity precise mathematical formulations. It was theorized that magnetism, light, and electricity propagated by means of electromagnetic waves. Radio waves were detected. Dmitry Mendeleyev devised the periodic table of elements based on the assumption of subatomic structures underlying chemical qualities. And the earth's geological past was probed in the rock studies of Charles Lyell.

Classical principles regarding the physical properties of matter and the mechanisms of their interactions, derived from Newton's seminal work, were also applied to the life sciences. In 1838 cell theory was formulated, revealing a basic structure of life at the microscopic level. This opened the way to the study of cell structure, where chemical principles were seen to govern the activities of cells and therefore all living matter. Taking this reductionist view a step farther, Jacques Loeb claimed that the instincts of lower animals are mere physiochemical reactions. At the macroscopic level Charles Darwin (1809–92) proposed a purely mechanical explanation, natural selection, for the evolution of animal and plant species. Here nature, operating by random mutations, mindlessly selects for survival those traits that adapt a species most successfully to unpredictably changing physical and biological environments. Mendel's theory of inheritance introduced the gene to science, which paved the way for the twentieth-century integration of natural selection and genetics.

These and many other scientific discoveries of the nineteenth century led to practical knowledge and inventions that rapidly transformed human life. Understanding the role of bacteria in disease ushered in dramatic advances in medicine. Pasteurization and the canning of foods soon followed. Chemistry opened the way for the internal combustion engine to power automobiles and eventually airplanes. Plastics, artificial fibers, and chemical fertilizers were developed. Steam trains and steam-powered, metal-hulled ships came on line, helped along by advances in metallurgy. Military technology expanded with the invention of the bullet cartridge, the revolver, the repeating rifle, the machine gun, and a wide range of new explosives. The exploration of electricity resulted in artificial lighting, the telegraph and telephone, electric generators and transformers, hydroelectric power, the battery, modern factories, and the recording of sound and motion pictures. It was as if some new god wearing a white coat had declared, "Let there be light!" and then thrown the switch.

THE DIVORCE

It was inevitable that the rise of science would have dramatic political and religious consequences. At this juncture, the mid-nineteenth century, a number of scientists emerged from their laboratories to crusade for political and religious changes that they felt were demanded by scientific knowledge. They became public speakers, writers, popular educators—salespeople for the science descended from Newton.

One of the most popular spokesmen for scientific materialism in the nineteenth century was the German physician and philosopher Ludwig Büchner. His book *Force and Matter* (1855) reduced the mind and consciousness to physical brain states produced by active matter. Büchner rejected religion, God, Creation, and free will, and in a later work denied there was any difference between mind and matter at all. In the same vein the Dutch physiologist and philosopher Jacob Moleschott expounded a theory that thoughts and emotions had a physiological basis. He became famous for the statement "No phosphorus, no thought."

The program of scientific materialism scored its greatest triumphs in England, largely due to the X Club. This informal group of nine men included the distinguished biologist T. H. Huxley, the philosopher Herbert Spencer, the physicist John Tyndale, and the botanist Joseph Dalton Hooker, all preeminent in their fields. Huxley in particular was a man of enormous talent and energy and a strong popularizer and defender of Darwin and the theory of evolution. In his scientific research he sought to explain physiochemical laws as the basis of living processes. In public he was a brilliant and pugnacious speaker who promoted science in public education and worked for the creation of a scientific elite. He trained schoolmasters in science, authored introductory science textbooks, held important posts in government, and was one of the founders of the journal *Nature*. Huxley claimed that science could achieve "domination over the whole realm of the intellect," and even spoke of the creation of a "church scientific."[4]

The philosophical bases for the group were provided by Herbert Spencer, another high-profile figure in Victorian England, who reduced social philosophy to scientific concepts. It was Spencer who coined the phrase "the survival of the fittest," associated since that time with Darwin's theory of evolution. These scientists-turned-promoters believed in a natural order to the universe determined by cause and effect, one that might prove unknowable, but one that was to be investigated by science, not religion. For the development of the intellect they thought that a scientific education was superior to the previous classical training (of classical literature, Latin, Greek, rhetoric, history, and moral philosophy). Members of the X Club held prominent positions enabling them to lobby successfully for official support of science and the teaching of science in all levels of education. Most importantly it was they who interpreted science to the public—a public that was becoming increasingly industrialized and secularized.

What did that interpretation boil down to? The five principles examined previously: objectivism, metaphysical realism, the closure principle, universalism, and physical reductionism. Yet the package sold to an unwitting public was labeled simply "science." Because scientific materialism had developed gradually and naturally from the interplay of theology and science, it is unlikely that these salespeople were even aware of the principles that guided their beliefs. By the nineteenth century these beliefs had become the scientific "gospel." Though unspoken, they were taken for granted as patently obvious. And although the roots of many of these beliefs could be traced to the Bible, their original cause had been forgotten, replaced, banished. God was now at most a ghost passively observing the machine that he had supposedly engineered with his "intelligent design." The object of scientific enquiry, originally the heavens created by God, had been replaced by "objective reality." God's sacred language, mathematics, had become a subset of the scientific method. The resulting universe could be likened to an immense clockwork, operating automatically—without morals or miracles—driven solely by the laws of nature. Humans and their thoughts and emotions were ruled by that machine. Scientists now saw themselves and the rest of humanity as organic robots.

The foundation of this mechanical philosophy was built with the stones and mortar of scientific materialism. Its tenets formed a grand metatheory, hidden behind the scenes, from which scientific laws were reasoned. These beliefs shaped the mentality from which science was interpreted. The images and explanations they conveyed to scientists and to the general public now seemed natural, intuitive, and rational. The mind's eye traveled the axes of Cartesian planes, plotted the proper angle for parallel parking, calculated the location, velocity, and spin of a tennis ball—whap! If x caused y and y caused z, x must be the cause of z! How could it be otherwise? The roots of modern-day secularism had been planted in the fertile soil of rapidly expanding scientific knowledge.

Scientific *What*?

Why is the distinction between science and this philosophy so little known today and what is its significance? *Scientific materialism* is an academic term for a science-based dogma, developed in Europe in the mid-nineteenth century primarily by German materialist philosophers (including Karl Marx) and their English counterparts—T. H. Huxley, the X Club, and others. These men certainly didn't all call themselves scientific materialists. Many of them were in fact eminent scientists. As we have seen, these science promoters firmly believed that religion was bankrupt as a useful guide to truth and that physical science held the answers to all important questions. So strong was their enthusiasm for an all-embracing scientific worldview that they often allowed their hopes, dreams, and beliefs to masquerade as facts.

They were especially impressed by Darwin's theory of natural selection. According to their own interpretation, natural selection meant that organisms best suited to win the competition for scarce resources survived, passing on their advantageous traits to succeeding generations—

Spencer's "survival of the fittest." This was a biological counterpart to the impersonal, clockwork universe of classical physics. For them, the no-nonsense, hard-nosed fact of this struggle for survival was the pattern behind every facet of life. Social philosophers influenced by scientific materialism created social Darwinism, the view that nations and individuals competed for economic supremacy in an arena where only the "favored races" or toughest individuals would succeed. There was no room here for any softness or idealism and, of course, such a philosophy gave at least tacit approval to war, imperialism, and racism. In like manner, Karl Marx reduced all aspects of culture to economics.

All of this was tied to "science." Marx called his philosophy "scientific socialism." By studying history "scientifically" Marx skimmed away utopian and romantic notions such as culture and consciousness, leaving the bare bones of economic struggle—capital versus labor. Similarly, by appropriating the name "Darwin" in its title, social Darwinism linked itself to the prestige of a leading scientist. It conveniently ignored the fact that Charles Darwin himself arrived at a softer version of evolutionary theory, one that factored in the changing natural environment and other elements and was thus not primarily concerned with a "competition" of the "fittest."

So nineteenth-century scientific materialists created a philosophy based on a set of beliefs that was not arrived at scientifically, or to put it differently, was supported by modes of inquiry that focused exclusively on material phenomena. They speculated beyond the scientific evidence into the realm of metaphysics, normally the sphere of religion and philosophy. Just as the public was unaware that there was more to this new philosophy than pure science, its promoters were themselves probably ignorant of the theological origins of the underlying tenets of scientific materialism—objectivism, metaphysical realism, the closure principle, universalism and physical reductionism. As we saw earlier, these elements had slipped into science surreptitiously beginning in the Renaissance.

According to these nineteenth-century crusaders, the message of science for human society was essentially this:

> Existence is purely physical—there is no other reality. The
> sources of this reality are the laws of nature, forces that are

entirely impersonal, having no connection whatsoever with the mind of human beings, their beliefs, or values. These laws operate in isolation from any supernatural, spiritual influences, all of which are illusory. Life in the universe is an accident, the outcome of mechanical interactions among complex patterns of matter and energy. The life of an individual, one's personal history, hopes and dreams, loves and hates, feelings, desires— everything—are the outcome of physical forces acting upon and within one's body. Death means the utter destruction of the individual and his or her consciousness, and this too is the destiny of all life in the universe—eventually it will disappear without a trace. In short, human beings live encapsulated within a vast, alien world, a universe entirely indifferent to their longings, unaware of their triumphs, mute to their suffering. Only by facing this reality and accepting it fully can humans live rationally.

Most people today, asked if this sounds familiar and where does it come from, would answer, "This is what science tells us about life and the universe." This is the philosophy of modernity, where, stated succinctly by the existential philosopher Jean-Paul Sartre, "everything is born without reason, prolongs itself out of weakness, and dies by chance." We have been exposed to this philosophy throughout our lives—in the classroom, in the media, by our doctors, and through the decisions of government agencies ruling on health, the environment, and elsewhere. It has been pounded into us consistently for so long that we've come to accept it as common sense. This, we are told, is what "non-believers" accept as the truth.

According to its proponents, this view alone is an authentic picture of the universe. Anything that deviates from it is (ironically) "metaphysical," idle speculation, or sheer fantasy. Perhaps one of the reasons for the strong polarity existing today between religion and secularism is the widespread influence of this view. In this modern "scientific" world, we are given a narrow choice: accept either scientific materialism or religious faith (which, according to scientific materialists, means turning your back on reality). For the strongly religious, this "message of science" is so radically opposed to their beliefs and traditions that it evokes anger, the entrenchment of religious fundamentalism, culture war, or jihad.

Wedded to classical science, scientific materialism shared in its enormous success, prestige, and influence, particularly in the nineteenth century. By the year 1900 most physicists believed that a complete understanding of the universe was only a few decades away. In the words of a supremely confident Lord Kelvin, "There is nothing new to be discovered in physics now . . ." All the great discoveries had already been made, leaving future generations of scientists with nothing more to do than to carry out calculations to the next decimal point of accuracy.

However, classical physics was about to take a fall. The heady optimism of the late nineteenth century had cost science a more open-minded, philosophical approach that had nurtured it in previous centuries (a facet we will examine later on). Overconfident and holding tightly to a dogmatic viewpoint, science was soon tripped up by a tiny detail—the atom.

When You Least Expect It 2
Materialism Undermined

MULTIPLE-CHOICE TEST

WHICH OF THE FOLLOWING statements best describes the atom?

1. The smallest component of the material universe
2. Unbreakable particles, invented by God to serve as the building blocks for the universe he created
3. A very small mass shaped like a round pudding. "Plums," in the form of electrons, are found on its surface
4. A nucleus made up of protons and neutrons, around which electrons orbit like planets circling the sun
5. A fuzzy, cloudlike nucleus surrounded by a cloud of electrons

The short answer is all of them and none of them. (1) The ancient Greeks theorized the existence of some "smallest component" of the material universe. (2) Early scientists in the West attributed these Greek building blocks to the work of their Christian God. Answers (3) through (5) describe attempts by science, beginning in the late nineteenth century, to account for a growing body of experimental evidence garnered from studies of subatomic particles—"atoms" that made up atoms. Answer (5), by the way, does not describe an atom composed of hard objects but views the atom as a grouping of probabilities. The atomic theories of the past may have satisfied their creators as being "true" mainly because they fit the cultural, religious, and philosophical background of the times. Physicists today, however, are just as uncertain about the nature of the atom—if there is such a thing—as we are today about many aspects of contemporary life. The more detailed story of the evolution of the atom that follows is a striking example of the instability of scientific theories, their dependence on broader cultural contexts, and their vulnerability to unexamined beliefs.

For science, matter holds a position similar to that of God in Christianity. Following in the footsteps of their ancient Greek predecessors, early scientists sought to reduce matter to its essential component, its lowest common denominator, the atom. If in the spiritual realm God was the first cause and the source of the highest knowledge and ultimate secrets of the universe, the atom fit that bill in the physical universe. Furthermore, if there were theological grounds for believing that the mind of humans could understand God, similarly scientists believed there was hope humans could understand the atom. Gradually the atom became a symbol for science's main quest, replacing the study of the heavens, astronomy, as its Holy Grail. This is still true, even though physics has advanced well beyond its early atomic theories. That the study of the elementary constituents of matter led to the most destructive power ever unleashed by humans tends to confirm, in a chilling manner, the atom's central importance.

ANSWER ONE: AN ANCIENT IDEA

Humankind has been searching for the atom, wondering if it exists, imagining what it might look like, for a very long time. Greek philosophers pondered the atom in the fifth century B.C.E., and we now know that it was proposed in India a century or two earlier. Leucippus first named it (*atomos* is Greek for "indivisible"), and his student Democritus developed a philosophy about it: within the void (an infinite vacuum), move an infinite quantity of particles making up the entire physical universe. These atoms are eternal, invisible, indivisible, infinitely small, and infinitely hard. Our impressions of matter—its temperature, color, texture, and so on—are imposed by our senses, whereas the atoms themselves are uniform, merely combining into different shapes.

ANSWER TWO: AN EDUCATED GUESS

Anyone can see that larger things naturally break down into smaller ones. When we crumble a piece of dry earth or crush rocks, or even when we mill wheat, we observe that these materials are made up of tiny particles. It doesn't take a great mental leap to reason that even smaller particles might be the prime components of the entire physical universe. This is a version of physical reductionism, one of the beliefs of scientific materialism we looked at earlier. In the words of the physicist Richard Feynman, "[T]here is nothing that living beings do that cannot be understood from the point of view that they are made of atoms acting according to the laws of physics."[1] If we also happen to believe that a creator such as God made everything, then knowing his basic material would help us understand the mind of God.

That was Newton's view. It was based mainly on his spiritual beliefs, and compared to the elegant empirical proofs of his physical laws, it was mere speculation. Newton's laws fit the physical universe on the large scale, the macro-universe, describing the motion of moving bodies that, compared to atoms, are relatively large and slow-moving. But he and his contemporaries, having no instruments to test their theory, could make only an educated guess that atoms operated according to those same laws. During the two and a half centuries between the Renaissance and 1900, scientists followed Newton's lead, assuming his laws applied equally well to all things big and small and refining them to a miraculous degree.

It was Newton's contemporary, Robert Boyle, whose study of air led him to propose the existence of atoms on experimental grounds. By observing the relationship between the pressure and volume of gases, he reasoned that atoms, as extremely small, indestructible constituents of matter, must exist. But Boyle's proposal was still largely intuitive. A rather large leap of faith was needed to connect his very basic exploration of gases to some invisible, atomic cause.

No really satisfying proof for the existence of atoms appeared before 1908, when Jean Perrin studied Brownian motion—the movements produced by water molecules when heated. By comparing the effects of gravity and Brownian motion on minerals dissolved in the water, Perrin was able to infer the mass (which is related to weight) of the particles causing this motion. Since that mass corresponded to the mass scientists had expected for atoms, voilà, the atom was real.

Keep in mind, though, that the atom began as a purely theoretical idea. This theory was based upon the way visible matter broke into smaller components. The idea that this was how things operated at a much smaller, invisible level was merely a guess—a guess that accommodated a belief among scientists that the universe was governed by underlying mechanisms we couldn't perceive. The atom was just such a transcendental idea. Afterward, through a variety of ingenious deductions—through reasoning—scientists became convinced of the atom's existence.

ANSWER THREE: THROUGH THE LOOKING GLASS

By 1908 one of the atom's basic properties, its indestructible unity as the smallest thing in the universe, had been literally blown apart. In 1897, the English physicist J. J. Thomson discovered a tiny, fast-moving particle with a negative electrical charge he called the "electron." If this formed part of it, the "indestructible and irreducible" atom had parts, subatomic parts. Since the atom as it was understood by science carried a neutral, not a negative charge, scientists reasoned that something else in the atom had to neutralize the electron's negative charge. Reaching for a visual representation of such an atom, Thomson proposed the "plum pudding" model, based on an earlier representation by Lord Kelvin. Electrons were embedded like plumbs into the surface of the as-yet-unknown positively charged "pudding" that offset the electron's charge, making the atom electrically neutral.

Proof of this pudding came in 1911, when Ernest Rutherford beamed tiny alpha particles (a form of radiation) into thin sheets of gold foil. These particles were so small compared with the size of atoms that they should have nearly all passed through the foil unmolested. However, Rutherford found that a significant number of them were deflected at various angles. In his own words, "It was almost as incredible as if you fired a 15-inch [artillery] shell at a piece of tissue paper, and it came back and hit you."[2] Something very hard in the gold atoms, harder than a swarm of lightweight electrons, must have been hit by some of his alpha particles, and this something he concluded must be the nucleus of the atom.

Through the observation of regularities, consistencies found in experimental data, Rutherford and the scientists who succeeded him contributed to a new model of the atom as a hard, compact nucleus surrounded by orbiting electrons—a solar model. All of this had been discovered by reasoning from scientific evidence. That reasoning was based on the underlying principles of classical Newtonian physics. If true, this was impressive knowledge for humans to have about a realm invisible to our senses.

ANSWER FOUR: THE SHELL GAME

But at this point a serious hitch developed. According to the classical laws of motion and electromagnetism, if charged particles such as electrons moved in circular orbits, they had to emit electromagnetic radiation. By doing so they would lose energy while orbiting and quickly spiral downward to crash into the nucleus. That couldn't be happening, because the building blocks of the universe were obviously more stable than that. No universe at all could be built upon such faulty, short-lived atoms. In 1913 the physicist Niels Bohr saved the day for the atom by theorizing that its electrons orbited in fixed locations. They were locked into invisible shells around the nucleus according to fixed energy levels, similar to the orbiting of the planets around the sun. A specific quantity of energy was needed to move an electron from one shell to another. They couldn't simply lose energy gradually and crash into the nucleus the way meteors fall to earth. So until something changed their energy, such as being struck by some other particle, electrons circling atoms were confined to specific orbits. This showed how the orbits and therefore the atom as a whole could remain stable.

Bohr's idea of shells with fixed energy levels came from a new science—quantum physics—which had been introduced to the world almost by accident in 1900 by the German physicist Max Planck. Planck had been trying to find a principle behind the way light radiation was emitted from materials when heated. According to classical physics, equal amounts of light energy should be emitted at all frequencies. But if that were so, since there are infinite possible frequencies, an infinite amount of light energy should be emitted from any heated object. Since it is unlikely that any material has infinite energy, that hypothesis was unacceptable. Planck discovered that if he modified several prevailing theories, he could derive a mathematical formula that fit the experimental evidence.

The principle behind it was *quanta* (Latin for "how much"). Light energy was emitted in quanta—energy bursts of fixed intensity and specific frequencies. Visualize a staircase with each step representing a quantum of energy. To move up or down to the next level of energy, you must go in single steps rather than moving gradually as on a ramp. This solved the problem by providing a theory that limited to realistic levels the amount of light energy released by heated materials. Planck's quanta gave Bohr what he needed to construct a model of fixed energy shells for the electrons of the atom.

Oops!

This model made sense of the atom, but largely wiped out Newton and company when it came to understanding the atomic and subatomic world. Consider just how serious this blow was for classical physics: its laws, aimed at obtaining a complete understanding of the "cause of causes" that scientists had been hunting for centuries, had proven inadequate. It appeared that the world of the tiniest, yet most basic and important constituents of the universe, might be governed by other, quite different laws.

Physicists took up this challenge with gusto. The three decades from 1900 to 1930 were among the most intense and fruitful in the history of physics. The quest to understand the dizzying implications of quantum physics (and relativity, which we will examine later) brought together some of the sharpest scientific minds on the planet. One of the most puzzling features—and one that must always be born in mind—is that quantum physics was strikingly successful in predicting the results of experiments. Furthermore, like the classical physics that preceded it, numerous important, practical inventions emerged from quantum theory, including lasers, transistors, computer chips, and photoelectric cells, not to mention the mixed blessings that came from understanding radioactivity: the atomic clock, atomic generation of electricity, radiation medicine, carbon-14 dating, and nuclear weapons.

But why is that success puzzling? It's because quantum physics does not in any normal sense explain how physical events happen. We are not given the kinds of pictures and logical descriptions we are accustomed to. The laws of quantum physics, relying heavily on mathematics, yield excellent results. They consistently conform to and accurately predict laboratory experiments at the atomic level. Many scientists consider quantum physics to be the most successful theory in modern science. But at the minimum, quantum physics implies that clear knowledge of physical processes as they occur in nature, independently of measurement, cannot be obtained. This was a severe blow to metaphysical realism: even if there are underlying, transcendental mechanisms governing matter and energy, according to quantum physics we cannot know them with any precision.

As we take a brief tour of the quantum world, we must leave behind many of our normal beliefs about reality. We need the open-minded curiosity of Alice in Lewis Carroll's classic tale *Alice Through the Looking Glass,* who said before the mirror, "Let's pretend the glass has got all soft like gauze, so that we can get through." Alice found everything on the other side read backward, and so shall we.

ANSWER FIVE: ON THE OTHER SIDE

In 1905 Albert Einstein made a key discovery about how light is emitted and absorbed according to quantum principles. His paper on the photoelectric effect (which earned him a Nobel Prize) demonstrated how a metal surface, when exposed to light or X-rays, emits electrons. In this process the light or X-rays, which had been traditionally viewed as waves, acted like particles. Einstein showed that this effect could be accounted for by using Planck's quanta concept: an atom of the metal could absorb one photon (light particle), but nothing less. That is to say that there was a lower limit below which the light would not affect the metal. Once that limit was reached, however, the metal emitted an electron instantaneously, as if a switch had been flipped. (Like one of those soft-drink machines that requires three quarters, not until that third coin drops does a can of soda emerge.) This was its quantum, the minimum "payment" in photons required for the metal to emit an electron.

Again, this ran contrary to classical physics, which viewed light as waves with smooth variations in intensity, like a dimmer switch. (Put in a penny, nickel, or dime and you get a few drops of soda. Add another coin and you get a bit more.)

So, challenging the general belief at the end of the nineteenth century that light was a wave phenomenon, quantum physics demonstrated the particle-like characteristics of light. It didn't stop there, however. It was soon shown that these particles, photons of light, could themselves act like waves. Keep in mind that waves and particles are viewed in classical physics as opposites. It was thought that waves of light or electromagnetism act according to laws that govern similar things, such as bodies of water. They reflect off surfaces in the same way waves do when they hit the side of a swimming pool. They interfere with each other, forming the patterns we see when we throw two stones into a placid pond. This is a physics of overall patterns. Classical physics applied the same math to waves of water, waves of light, and electromagnetism.

Particles, on the other hand, act as individuals. Instead of overall patterns sweeping along in smooth curves, they manifest as distinct, individual activities. Billiard balls do not move about like waves sloshing on the surface of a pool table, they are knocked about one by one. However, quantum physicists theorized, and later proved experimentally, that at the atomic level, particles—photons, electrons, neutrons, protons, and other subatomic entities—could also behave as waves, creating reflection, diffraction (bending around obstacles), and interference patterns. And they could do so not just through interactions with other particles but by interfering with *themselves*. Try to imagine that! A single particle could interfere with itself as if it were a wave.

These surprising results led physicists to theorize that subatomic phenomena could not be pigeonholed as either particles or waves. At the atomic level, at least, things somehow had both characteristics. Furthermore, these dual aspects mysteriously complemented each other. This became known as the "wave/particle duality." The principle involved is called *complementarity*. In the words of Niels Bohr, who coined the term, the two complementary aspects were not contradictory but "only together offer a natural generalization of the classical mode of description."[3] In other words the only way one can make heads or tails of this behavior is to combine the two opposite activities—wave/particle.

AND IT DOESN'T STOP THERE . . .

If this sounds bizarre, you are catching on. The wave/particle duality simply cannot be pictured in our normal, classical way of viewing things. But there was more to come. As physicists continued their explorations of quantum phenomena, they discovered that certain dual properties that are required to accurately describe physical reality could not be measured together at the atomic level.

In the game of baseball the pitcher on the mound at the center of the diamond throws the ball in a relatively straight line over a specific distance to home plate. If the batter hits a home run to center field, we can give an accurate account of what happened to the baseball in the two or three seconds it took to travel from the pitcher's hand to the batter to the center-field bleachers. We can measure the ball's mass and velocity—how long it took to get to the bat, how long it took to reach the bleachers. Then, without too much difficulty we can calculate the ball's velocity and location, and therefore its trajectory, at every point along its journey. From Newton's laws of motion we get a complete picture of events.

Quantum theory, on the other hand, predicted, and experiments have proven, that at the atomic level both the momentum (mass times velocity) and the position of a particle (such as an electron circling an atomic nucleus) cannot be measured at the same time. If you get an accurate measurement of the electron's momentum, you cannot obtain accurate information of its location. Conversely, if you closely measure the location, you cannot tell its momentum. Therefore you cannot obtain its trajectory. You cannot get a complete, classical scientific (or common sense) description of its motion. The same goes for other subatomic particles and for other paired attributes, such as wave/particle measurements, energy of a system in relation to time of system change, and the path-versus-interference pattern of things in motion. This lends physical reality a disquieting fuzziness. One may wonder, if we cannot measure the characteristics that traditionally account for the reality of a phenomenon, does that phenomenon as we normally think of it actually exist? Or could "it" be something as yet unknown to us? We will see in part 2 that some Eastern philosophies have developed sophisticated explanations and descriptions regarding such questions of existence and nonexistence.

ARE YOU SURE? THE THEORY OF UNCERTAINTY

This brings us to another of the pillars of quantum physics: uncertainty. According to this principle, the location and momentum of a particle such as an electron circling an atomic nucleus exist only as probabilities. These probabilities can be discovered mathematically, but they are only highly educated guesses. This corresponds to the fifth possibility in our atomic multiple-choice test. The electrons spinning about the nucleus, as well as the protons and neutrons that comprise it, might as well be visualized as clouds of possible locations. Now, that really shakes our traditional view of matter as something solid. Prior to the quantum revolution it was already known to science that the atom, the building block of the physical universe, was mostly empty space, because the electrons orbited at relatively large distances from the tiny nucleus. But along comes quantum physics and even that spaced-out model fades into a ghostly fog of uncertainty.

The theory of uncertainty, however, is not some vague philosophical abstraction. First formulated by the Nobel laureate Werner Heisenberg, it has its own specific mathematical formulation that can be applied successfully for predicting the outcome of laboratory experiments. At the same time, it strikes at the very heart of scientific materialism, of the whole realist stance of traditional science, and at the popular, common-sense view of reality. Classical physics is based on the proposition that if you can measure the initial conditions of any physical system, then by discovering the forces acting on that system and applying the laws of classical physics, the activities of that system can be precisely known. Quantum uncertainty states that on the atomic level the initial conditions, if such things exist, cannot be precisely known at all. We cannot even begin.

Another surprising fact about the atomic world brought to light by quantum physics involves measurement itself. Suppose we wish to make some kind of direct measurement of an atom. In classical physics, measurement is an entirely passive act. It is assumed that the object to be measured has real properties and that all we have to do then is examine or measure the object in order to discover those properties. Of course, to measure some atomic property, an instrument must interact with the atom. This could be the light used in a microscope or the alpha particles Rutherford used to divine the existence of the atomic nucleus. In classical physics, disturbances could be satisfactorily reduced by gradually lowering the intensity of the effects of the measuring device or by compensating for these effects. But quantum physics has shown that when we use an instrument to measure an object at the atomic level, the device dramatically and unpredictably disturbs the object. If you literally throw some light on our atom, it will be instantly knocked askew. The photons (or waves) of light strike the atom with comparative violence, creating an immediate and unpredictable reaction. Moreover, since light cannot be dimmed gradually to a suitably low frequency or intensity (because, as we saw earlier, it is emitted in steplike quantum jumps of energy), there is no way to "objectively" measure the object. It would be like trying to sort beads with a fire hose. This problem has been demonstrated both mathematically and in practice. Furthermore, if the influences of measurement were unpredictable, how could we compensate for them?

IMPOSSIBLE, YET PROBABLE

But the problem of measurement runs even deeper than that, taking us to realms even Lewis Carroll could not imagine. In a variety of quantum experiments, it has been shown that the very notion of an atomic object having a fixed, preexisting state is doubtful. In classical physics, uncertainty regarding a measurement means that we simply don't know yet what the initial, fixed state of a real object is. Either we haven't yet made the measurement or we haven't yet figured out a way to do so. We are simply in ignorance. In the quantum world, the mathematically calculated possibilities for a future measurement exist as a remarkable state called a *superposition*.

To give a crude example, suppose we wish to measure the exact point at which a bullet strikes a wall. In this example a person points a rifle, firing a single bullet in the general direction of a concrete wall. The shooter will not sight down the barrel to aim the shot. The bullet could hit any part of the wall. Let's say that the bullet will travel for a tenth of a second from its place in the rifle's firing chamber to the spot where it strikes the wall. From a classical-physics (and commonsense) point of view, the possible locations for the bullet's impact are a mixture of all the locations on the wall. It has an equal chance of hitting any one spot. When it hits, that constitutes a measurement and we are no longer in ignorance as to the bullet's trajectory.

But quantum physics views this in a completely different manner. In that tenth of a second between the firing of the rifle and the impact of the bullet on the wall, the bullet is conceived of as existing at every possible point between the rifle and the wall. It's as if the rifle fired a cloud of bullets as wide as the wall. Mathematically this is formulated as a wave function of probability represented by a graph with accompanying mathematical formulas. When the bullet actually hits the wall at a specific spot, it is said that the wave function "collapses" to a single point. One might complain that this collapse of the wave function is merely a fanciful way of describing the commonsense point of view: we are ignorant of the bullet's trajectory, but when it hits the wall, that is essentially a measurement—we have discovered its trajectory.

However, due to numerous experimental results, most physicists hold that a subatomic "bullet," such as a photon of light fired at a "wall" of light-sensitive film, exists not as a mixture of chance possibilities but as a superposition. The photon, in some mysterious manner, "exists" as a quantum system covering all possible positions prior to the collapse of the wave of probability to a single point. The photon is *everywhere* prior to its striking the film. The classical concept of our "ignorance" of the photon's true trajectory is replaced by the quantum concept of *uncertainty,* wherein the photon has no single trajectory in terms of a well-defined state.[4]

IMPOSSIBLE, YET TRUE

Does all of this sound like magic? Have these quantum scientists let their imaginations run amok? Perhaps. Let's take a more practical example—

atomic radiation. We can all see its results in medical X-rays, or the glowing dials of clocks and watches (which use radium paint). The concept is simple: Radiation results from changes within an atom that release particles. Radioactive substances, such as radium, decay naturally by releasing these particles. Radium emits particles that are visible to us as glowing green light. However, from a classical standpoint this is impossible. The light you see from the radium painted on clock faces shouldn't be there. Why? The attraction force of the nucleus of such substances is stronger than the kinetic energy of the escaping particles produced from decay. The particles should be trapped in the atom with no more chance of escaping than any of us would have escaping earth's gravity by jumping into the air. Yet quantum physics provides a mathematical explanation for this very common occurrence (it's called tunneling) and it does so without telling us how it happens in classical, commonsense terms. Yet, it does happen.

At this point you might ask, "If this quantum craziness predominates at the atomic level, why not in the visible world? Isn't the universe as a whole made up entirely of these microparticles?" In fact quantum laws do affect the macroworld. Their effects can be calculated just as easily as for the microworld. From a quantum physics point of view, classical laws merely provide a close generalization at the macrolevel of the more accurate quantum measurements. At the level of gross matter they are equally accurate. But at the atomic level many classical laws fail. A yardstick will give you a reasonably accurate measurement of the length of an aircraft carrier, but is useless for measuring the size of molecules. An electron microscope could measure both.

OPPOSED TO GAMBLING

Despite his role in the development of quantum physics Albert Einstein was far from happy with it. "God doesn't play dice!" was his reaction to quantum uncertainty.[5] Much of the last thirty years of his life was spent trying to find a classical, realistic basis underlying the proven and accurate predictions of quantum physics. For Einstein, because it couldn't show precisely what was going on, quantum physics was "incomplete."

Beginning in 1927, Einstein devised a series of challenges to quantum physics, most of them aimed at Niels Bohr, its strongest defender. Since the experimental means were not available to test them at the time, Einstein's doubts were formulated as thought experiments dealing with quantum principles. Later, with improvements in technology, Einstein's challenges were tested. In each case the predictions of quantum physics were confirmed.

As we saw earlier, quantum physics says that measurement disturbs the objects being measured so much that a complete description of a two-part, complementary system such as the momentum and position of a subatomic particle is not possible. A clear picture doesn't emerge. In 1935 Einstein and two associates (Boris Podolsky and Nathan Rosen) thought they had found a flaw in this uncertainty. What if a *single* measurement could reveal two complementary aspects of such a system? They proposed a scenario where a pair of protons (subatomic particles that help make up the nucleus of an atom) was shot from a device in opposite directions simultaneously. Imagine two cannons placed back to back. A single trigger fires a projectile from each cannon at the same time but in opposite directions, each ball spinning in a different direction as it emerges.

In the case of the proton pairs, it is known that each must have an opposite spin. (Protons along with other subatomic particles have what is called a spin angular momentum, something like the spin on a tennis ball.) In an experiment set up in this way, proton spins can have one of only two values, designated as "$+\frac{1}{2}$" or "$-\frac{1}{2}$." These opposite spins are the only alternatives. So here the two protons with opposite spins would form the complementary system to be measured.

Again, in such a paired system if the spin of one proton is $+\frac{1}{2}$, the spin of its twin must be the opposite (spin $-\frac{1}{2}$) and vice versa. Therefore by measuring the spin of one twin, the spin of the other would be known automatically—without a second measurement being necessary. One measurement would give complete knowledge of the two-proton system—the clear picture classical physics and Einstein demanded. This would be equivalent to learning the position and momentum of a subatomic particle with only one measurement rather than the normal two. Quantum physics denies the possibility of measuring both of these attributes simultaneously by any means. The possible spins—$+\frac{1}{2}$, $-\frac{1}{2}$ for one proton, and –½, +½ for the other—are viewed by quantum physics as a superposition of all of these possibilities, prior to measurement.

Einstein, Podolsky, and Rosen (or EPR) believed that the complete two-proton system could be measured for spin because at the moment the twin protons were emitted their spin was already determined. It was just a preexisting quality of the proton pairs. Nothing changed once the protons were emitted. The measurement would be entirely passive. After the two protons were emitted from the device, all one needed was to measure one proton. It was obvious to them that the measurement of one already-existing spin revealed the already-existing (opposite) spin of the other proton, yielding complete understanding of the system. This completeness was the way classical physics worked, and anything less than that was incomplete knowledge. The alternative explanation given by quantum physics was that by measuring one of the pairs the system was disturbed, altered. Here, by measuring one of the protons, the other was forced to have the opposite spin. Therefore the measurement somehow created the spin state for the system (and might do so over an arbitrarily long distance), or (*gasp!*) there was no original, specific state of the system prior to measurement! Those alternatives were unacceptable to EPR, so quantum physics must be flawed. Or so they thought.

This thought experiment, known as the EPR paradox, stumped Bohr and remained troublesome for quantum physics for thirty years. Nevertheless its final solution, theoretically in the 1960s and later through experiments, provided one of the strongest challenges imaginable to classical physics. If quantum physics says that we can accurately measure only one of two complementary aspects of a system, the second, unknown quantity can be called its hidden variable. (In the formula $1 + x = 3$, the hidden variable is simply x.) In Einstein's way of thinking, if the spin of one proton were +½, the hidden variable would be the other spin, which in this case would have to be -½.

In 1964 and 1966 the physicist John Stewart Bell published two theories dealing in very general terms with hidden variables. Bell's theorems (called Bell's inequalities) proved on paper that no hidden-variable theory based on classical principles could match the accuracy of quantum physics.[6]

Bell's theorems were given definitive laboratory tests beginning in the

1980s. These experiments showed that the EPR assumptions challenging quantum physics were incorrect. Einstein and company were wrong. It was demonstrated that measuring the spin of one of the proton pairs did affect the spin of the other. Furthermore, it doesn't matter whether the measurement of the protons is made in a single laboratory or on planets light-years apart. Measurement itself plays an active role in the system, which must be viewed as a set of dynamic interactions whose scope is not limited to a specific locality.

Classical hidden-variable theories presume three things: (1) locality (things done at one location have effects only at that locality) (2) strict causation (some event or sequence of events—causes—are responsible for observed effects), and (3) reality (phenomena have real properties, not fuzzy probabilities). But Bell's inequalities and subsequent experiments proved that information—in this case measurement—plays an active role in the system. Indeed, the distances over which such effects occur expand our normal notions of what constitutes a "system." This is called quantum nonlocality or entanglement. But as always in the quantum universe, a clear picture of how it happens is not possible. And if there can be no complete knowledge of the system or phenomena, strict causation and reality are also left hanging. So in this case it appears that Einstein's complaint that quantum physics was incomplete ironically led to proof that it is the classical view that is inadequate.

Incidentally, this mutual entanglement of phenomena has practical applications. One is in the creation of hyperspeed quantum computers. In 2004 physicists working in this direction reported observing paired photons able to "communicate" over a distance of a third of a mile. The mere observation or measurement of one of a pair of particles in an entangled state affects the other particle in the manner demonstrated above in the EPR paradox. Theoretically, particles in different galaxies may be able instantly to affect each other in similar fashion. Recently, a team headed by Anton Zeilinger of the University of Vienna pioneered the "one-way quantum computer," where the observation of entangled quantum bits drives the calculation. They were able to make a calculation in a record 150 nanoseconds (150 billionths of a second).[7] Thus, entanglement has opened up some exciting possibilities for new technology. Furthermore, it has features that parallel theories of the interdependence of mind and matter found in contemplative traditions—theories we will explore in detail in part 2.

RELATIVITY

The appearance of quantum physics at the beginning of the twentieth century was only one punch in a brutal combination to the chin of Newton's universe. Albert Einstein introduced relativity in two stages: a special theory of relativity (how basic phenomena such as time, velocity, matter, and energy are relative and not absolute in nature) in 1905, and his general theory of relativity (on gravity as curved space-time) in 1916. In the classical universe, time and space are absolute, the same everywhere. Accordingly, if astronauts were able to travel at the speed of light to a point in space twenty light-years distant from the earth, the round trip would take forty years. The clocks and calendars in their spaceship and those on earth would be perfectly in synch. When they returned to earth, all of us would be forty years older, including the astronauts.

But according to relativity, there would be a difference in the way time passed for us on earth and for our star voyagers. Here the relative motion between bodies alters not only time but size. Although the astronauts' clocks and calendars wouldn't pass any more slowly for them, they would when compared with clocks and calendars on earth. If there were some way for us on earth to observe the clocks aboard the spaceship, they would be seen to be moving more slowly than ours. Objects on board, including the rocket itself, would appear shorter. And if our astronauts were traveling at 99 percent of the speed of light to a point twenty light-years from earth, when they returned, we would be forty years older, but they would have aged only 5.7 years. At that speed, time would have slowed seven times in comparison to a relatively fixed position such as that of earth.

These changes in time and size happen at all relative speeds but are only noticeable at velocities nearing the speed of light (186,000 miles per second when traveling in a vacuum). This effect has been proven in the laboratory and is important in the use of particle accelerators to explore the subatomic world. The particle accelerator speeds up subatomic particles to velocities very close to the speed of light. At such high speeds and energies, the particles can bombard objects or other particles creating measurable effects from which physicists can draw theoretical conclusions. The particles under observation are incredibly tiny and short-lived, very difficult to catch sight of with scientific instruments. The difference in speeds between the particle and the observing scientists causes time dilation: just as clocks on our hypothetical spaceship are slow in relation to an outside observer, the quick decay time of these exotic particles slows slightly for an observer in another frame of reference, the relatively stationary laboratory. That dilation of time is just enough to allow the fast-moving particles to register on detection devices.

One of the keys to special relativity is the speed of light, believed to be the fastest velocity obtainable, and a constant. The speed of light is the one thing that does not change for different observers. It's the c in the famous formula $E = mc^2$. That revolutionary equation showed how energy and mass are interchangeable; they are different forms of the same thing. It also suggested that there is enormous energy wrapped up in the force holding the atomic nucleus together, since to determine the potential energy in an object, you are multiplying that mass by 186,000 squared—about 34.6 billion! This relationship was one of the keys to understanding atomic fission, which led to the powerful chain reactions of nuclear reactors and atomic weapons.

The general theory of relativity presents a completely different view of gravity from that of Newton. Its foundation is not just space but, as we saw earlier, space-time—the two are a unity. According to this theory, space-time is the stage whereon the entire drama of physical phenomena is enacted. As we touched on earlier, for relativity the gravity exhibited by physical objects in space does not "attract" other objects. Rather, objects create a curvature in the space-time continuum in their vicinity—the more massive the object, the greater this curvature in the fabric of space-time and the wider its reach. The uniform dimensions of Euclidean geometry, with its straight lines and planar geometrical figures, are replaced by patterns sometimes likened to the shape of a lumpy mattress. Moving objects follow these curves. For example, the space-time curvature caused by the mass of the earth forces nearby moving objects such as meteorites to bend in our direction. Following the natural contours of space-time, such objects move toward the earth because that's where the "road" bends. This makes gravity a product of the uneven shape of space-time itself, rather than a force of attraction, as Newton thought. Again, just as the majority of scientists once took Newton's view of gravity as gospel, today Einstein's completely different theory predominates.

BACK TO THE DRAWING BOARD

By the 1930s it was clear to physicists that the atom, in its three-part model (clouds of electrons, neutrons, and protons) was no longer adequate to describe the physical universe. The theoretical implications of quantum physics and relativity, along with new, more sensitive laboratory instruments, were rapidly complicating the atomic picture. A swarm of new particles including "virtual particles" (said to perform some necessary theoretical function but not observable) were required to make sense of the whole. As they were needed, most of these new entities dutifully appeared in laboratory experiments. Antimatter is one example. First theorized in the 1920s as necessary for an understanding of particles' electrical charges, the existence of antimatter was confirmed in laboratory experiments in 1932. This discovery led to an understanding that all subatomic particles must have matching antiparticles, thus doubling the number of known particles and adding a whole new dimension to atomic physics. However, the confirmation of these particles in the laboratory involves very specific conditions. They are created with high-energy bombardment from other particles and are extremely short-lived. (The anti-atom, created in the laboratory in 1995, lasts a mere 40 billionths of a second.)

In addition to all these new particles clouding the picture was the problem that quantum physics, relativity, and those classical concepts that remained useful at the subatomic level often didn't jibe with one another. Therefore gradually a compromise was hammered together called the *Standard Model*. The following description is presented merely to give you an idea of how rich and complicated the subatomic scene has become:

In brief, the Standard Model envisions the subatomic world as divided into two major types of particles and four forces. The forces are gravity, electromagnetism, strong nuclear force, and weak nuclear force. Of course, gravity and electromagnetism had been with physics for a long time. The strong nuclear force, responsible for the enormous release of energy in nuclear fission (where atoms break apart), is basically the binding power of the atomic nucleus, the force holding protons and neutrons together. The weak nuclear force is responsible for the process of decay that can turn a neutron into a proton. The nuclear reactions that cause the stars to burn are controlled by this weak nuclear force.

There are now over two hundred subatomic particles catalogued; however, the so-called fundamental subatomic particles (which are thought to be point-like) are the *quarks* and *leptons*. The quarks, if they carry electric charge, participate in all four forces of nature and are the major constituents of the protons and neutrons that form the nucleus of the atom. Leptons do not participate in the strong force, but otherwise participate in all the other three forces if they happen to carry electric charge. Our old friend the electron is an example of a lepton. According to the Standard Model, all the varied behaviors of subatomic particles can be adequately described using only six quarks, six leptons, and their antiparticles. The agent of action between them is the *gauge boson*. This is a messenger that transmits the forces among particles. The photon, for example, is the gauge boson of electromagnetism. Keep in mind that this is a simplified version of the Standard Model. The relationships among these particles are complex and, like many of the particles and their attributes, highly theoretical.

You may be forgiven if you are left with the feeling that the denizens of the microworld and their complicated goings-on are similar to the epicycles of Ptolemy's solar system or the promises of politicians: a little overextended to be quite believable.

THE LATEST

Before we sift through the ashes of classical physics to see what has survived, let's briefly review some of the current theories of physics—narratives that would challenge Lewis Carroll or any other writer. The Standard Model, with all its faults, is still the main working hypothesis of physics and the bread and butter of undergraduate physics programs. But there are several promising and quite novel approaches that have caused excitement in recent decades.

Chaos theory, with its origins in the late nineteenth century, was an important early challenge to the classical view of the universe as orderly clockwork. It is now a major scientific discipline (within the field of dynamical systems theory) embracing several scientific areas. The underlying concept is (long-term) unpredictability. Chaos theory helps us to understand random behavior as found in turbulence, weather, ecology, fluid motion, acoustics, electrochemistry, and even disease epidemics. For even simple systems of three or more interacting particles, the system evolves in such a manner that choosing almost identical initial conditions can led to completely different (exponentially divergent) behavior at a later time. This sensitivity to initial conditions means that to obtain long-term predictability, it is necessary to make, at some initial time, measurements that have exponentially small errors. However, this turns out to be a practical impossibility even for a system with three particles, let alone a gas, such as the atmosphere, or a liquid, such as the ocean.

One tangible result of chaos theory has been fractal geometry. Fractals are shapes made up of smaller, scaled-down versions of the larger figure. Some snowflakes are fractals: the flake is composed of smaller identical flakes. The mathematical relationships determining fractals have resulted in a science permitting some understanding of irregular heartbeats, fluid turbulence, the motions of star clusters, and population dynamics, among other applications. Fractals have helped revolutionize computer graphics and help mapmakers to model irregular shapes such as mountain ranges and coastlines. So not only is the universe apparently more irregular than once supposed, but this irregularity can be understood and put to practical use.

Perhaps the most promising new research in physics over the past two decades has come from *superstring theory*. One of its predecessors, string theory, was devised in the 1970s to describe the strong nuclear force. Quantum chromodynamics—a theory developed to describe this force acting among subatomic particles—proved more efficient for that task, but in the 1980s string theory was combined with another new concept, supersymmetry, thus creating the superstring approach. Whereas the Standard Model treats elementary particles as points in space, superstring theory thinks of them as one-dimensional, vibrating, stringlike entities. If this is rather hard to visualize, don't worry. Superstring theorists take this description with a grain of salt, deferring to the highly complex mathematics that describe it (to the few who can understand its exceedingly difficult formulations). What excites physicists about superstring theory is its treatment of gravity. The Standard Model has been unable to integrate gravity with the other three forces to provide a self-consistent, unified quantum theory of the atomic world. Superstrings may be able to accomplish that by bridging the gap between Einstein's gravity of curved space (which is a branch of classical and not quantum physics) and quantum physics (to which superstrings relate).[8] However, superstring theory also presents some knotty problems. It depicts a universe of eleven or even twenty-six space-time dimensions instead of the normal three. It cannot explain the masses of known particles and suggests the existence of shadowy new particles whose attributes defy current understanding.

Another important area of research, and one that has deep philosophical implications, is the physics of the vacuum. In classical physics the vacuum of space was considered empty, merely a three-dimensional realm in which the important stuff, matter and energy, manifested. But modern research has questioned that. One view proposes that the vacuum of empty space—with everything we know of in terms of energy and matter removed—has infinite energy. An alternative explanation views the vacuum as having zero energy but still possessing structure and the potential for creating phenomena. Many physicists theorize that a fluctuation in the energy of the vacuum was the source of the big bang. As we shall see later, current theories of the vacuum offer provocative suggestions as to the origins of the universe, with striking parallels to the deepest dimensions of consciousness.

A Conclusion: The Falling Dominos of Scientific Materialism

Superstrings are but one new and promising candidate for something physicists have sought since the beginning: a TOE (theory of everything) or GUT (grand unified theory). That is what scientists at the end of the nineteenth century thought they had accomplished—a complete understanding of the physical universe derived from Newton's fundamental laws. Today, however, the picture is much more complicated. There are far more theories and experimental data and no one is sure to what extent they correspond to reality or even if seeking reality is the point. Given past failures and the proliferating implications of relativity and quantum physics, one wonders if the complexities of even the accepted Standard Model can hold water. By seeking grand unified theories that promise something *real,* is physics producing yet another complicated house of cards doomed to collapse from its own weight?

Whether you call it a theory of everything, the atom, or the "mind of God," when the quest of science is defined in such terms, the game is objectivism. That's the age-old desire for a comforting, freestanding, *transcendental* universe acting behind the scenes. However, can one find such a mechanism somehow isolated from the human mind and the human condition? What do the discoveries of atomic physics say?

If we are after some essential physical object—matter—the theory of relativity alone would place us on shaky ground. The matter we are seeking cannot be the solid stuff Newton conceived of. As we have seen, depending on the position and velocity of the observer, matter varies in mass, size, and duration. This applies equally to mountains, molecules, and subatomic particles. The change everything is constantly undergoing may be too small for us to perceive in terms of a mountain, but according to physics, that mountain is made up of subatomic particles, and according to relativity, their mass, size, and duration are not fixed. After all there is no point of absolute rest anywhere, is there? A particle acceleration laboratory in, say, Berkeley, California, is itself moving at perhaps 500 mph as the earth spins. The earth is moving around the sun, and the sun is moving in relation to stars and galaxies. If current theories of the big bang are correct, everything in the universe is moving away from everything else and that motion changes matter. So whether we are measuring effects on a large scale for an object with slower speeds or we are measuring the rapid velocities of that object's subatomic components, matter, according to relativity, is very slippery stuff.

Matter is undermined all the more by quantum physics. Do subatomic particles exist as waves, or particles, or some indescribable "complementarity" thing? Recall that Niels Bohr used the term *comple-mentarity* merely in an attempt to provide a "generalization of the classical mode of description"—a useful picture. Although the formulas of complementarity can account for the bizarre behavior of subatomic particles, they don't really describe it. What kind of reality can only be partially measured—that is, you are allowed to know its momentum but not its position, and vice versa? ("The express from Washington to New Haven is moving at 72.5216 mph, but we don't know exactly where it is right now.") What kind of matter exists as a superposition—covering all possibilities before it is measured? ("The train and its passengers are everywhere between Washington and New Haven, but we won't know anything exactly until the train stops, allowing its wave function of probability to collapse so we can check it out.")

How can we picture a micro-universe, the world supposedly underlying everything we perceive, if trying to observe it closely—measuring it—makes it unobservable, changes it, or somehow participates in its creation? If our measurements are a participatory act and cause the universe to change, there can be no absolute objectivity. Rather, phenomena become part of a dynamic web of interactions. Furthermore, if causes and effects are nonlocal, could subatomic activities in some other galaxy be intimately entangled with events here, and not over long periods of times but instantly? Thus, objectivism and physical reductionism have been severely questioned by the conclusions of modern physics.

Recall that another tenet of scientific materialism, the closure principle, closed off the physical universe from nonphysical influences such as gods, demons, and the human mind. Yet in quantum wave/particle experiments it is the choice of the scientific mind to seek either a wave or a particle that determines the outcome. In the EPR paradox (which suggests all phenomena may be entangled), subatomic measurements, superpositions, and elsewhere, it is *information* that determines the physical system. The very characteristics that make data "informative" are determined by the human mind. It is we who determine which data are relevant information and which are only background noise. This alone suggests that the mind is an intimate participant in physical events; it is not closed off from them. Here science is approaching an all-embracing view of mind, which—as we will see in part 2—has a long and influential history in Eastern philosophies.

Is the reality of the universe to be found, then, in principles and laws underlying or transcending its questionable material presence? This would be transcendental realism. Yet what kind of reality do the mathematical formulations of atomic physics have? In relativity and quantum physics the math works, but for what? It works for kinds of "matter" that don't at all conform to our usual notions. If mathematics or concepts such as curved space-time are the realities generating our universe, they imply a universe that is a mere ghost of what we normally assume it to be. What does $2 + 2 = 4$ mean if there are no objects to count?

Beyond that, as we have seen, the abstract theories of contemporary physics are many and they are not without both external and internal inconsistencies. This is not surprising, because the procedures of science today often do not and cannot follow the niceties of time-tested scientific methods. Rather, the data and theories revealed over at least the past century have forced researchers to rely on a more improvised, pragmatic approach called "bootstrapping" (from the old adage "pulling yourself up by your bootstraps"). With so many theoretical possibilities that might guide research—relativity, quantum physics, chaos theory, super-strings, and others—physicists depend to a large degree on "whatever works." Such pragmatism has produced results in no less abundance than in earlier days, when classical principles allowed for more orderly and systematic scientific methods.

However, if pragmatism plays a large part in research, where does that leave transcendental realism? Where are the orderly, universal laws behind the veil of appearances if our best theories (and their mathematics) have no all-embracing order? Even mathematics itself—Galileo's divine language—can no longer be viewed as a clean, consistent, homogenous group of truths or axioms. It has been proven that the underlying axioms for any single mathematical system—even arithmetic—cannot make that particular system complete and consistent. Axioms are simply conventions. Moreover, different mathematics, like checkers and chess, each have their own rules and applications. There is thus no one, true mathematics.[9]

This presentation of the history of physics has by necessity been very general and gives a mainstream view, but there are other interpretations. In fact viable classical replacements have been found for some quantum solutions (although not to the extent that these threaten quantum physics as a whole). Classical concepts have sometimes been successfully combined with quantum physics and relativity in specific situations. At the cutting edge of physics, scientific materialism may be in tatters, but many scientists still argue that the rug has not been completely pulled out from under simple realism.[10] Can scientific theories adequately describe "reality"? Or does the nature of things deny that possibility? Though only time and continued scientific investigation will tell, it is fair to say that contemporary physics has cast some very serious doubts on realist positions.

Perhaps we can find greater clarity by taking a deeper look at how science operates. What is going on behind the veil of scientific theorizing and experimentation? What about the mind that is doing the theorizing and creating the experiments? Where does it fit in—and for that matter, what is the mind? Returning to our multiple-choice test, "Which statement best describes the atom?"

When Believing Is Seeing 3
How Belief Influences Perception

IMAGINE you have arranged a meeting with someone special. This could be a first date with a person to whom you are strongly attracted or it could be a casual, one-on-one business lunch with an important prospective client. Perhaps you're having dinner with someone you greatly admire—a political or spiritual leader or an artist. Or you may be expecting an attractive job offer.

Now let's suppose you will meet this person in a familiar café. Feeling a tingle of excitement, you arrive early. It may be the same old café, but it looks cheerier than usual. Your senses are keen, you are more aware of the chitchat going on around you, the colors are more vivid, and there is a welcoming air about the place. The waiter throws you a friendly greeting as he glides by. You are in excellent spirits, in a mood of high expectation.

Soon the appointed hour arrives, but where is your guest? The minutes tick by, but the person you are expecting doesn't show up. You glance at your watch and see that it is now ten minutes past the agreed meeting time, then fifteen, then twenty. "Should I give a call?" you wonder. Your date is now a half hour late and your buoyant mood has evaporated. You are annoyed when the waiter asks a second time if you wish to order. Feeling warm and sweaty, you shift from side to side, take some deep breaths. The café has taken on a dark, oppressive appearance. Just then your guest strides through the door, saying, "Sorry, I got stuck in traffic!" Immediately a sense of relief floods your body and a big smile creases your face. And the glow returns to your surroundings.

We've all had similar experiences. However, although they are common, we rarely step back and ask ourselves, "How and why do my expectations alter my perceptions?" One moment the café seemed extremely pleasant, then it became progressively duller, perhaps even depressing. Then suddenly, in an instant, it became vibrant and exciting again. Here, then, is a case where "believing is seeing." Our six senses, so important to our knowing what's what, are vulnerable to our expectations and the moods they generate. Our expectations, in turn, are based upon our beliefs. In the scenario just described, we believed our date would be on time. When it appeared we might be stood up, the mood shifted. The mere confirming or dampening of our expectations can alter our perceptions.

A similar phenomenon occurs frequently in eyewitness accounts, which can vary greatly from person to person. Observers see things differently due to temperament, moods, and other predispositions—differences in how their memories and even their senses operate. Just think of the biases that so clearly come into play when coaches, players, and fans argue with the decisions of a baseball umpire. Furthermore, one's beliefs can have a particularly dramatic impact on perception when we are looking at something we have never seen before. When an American fleet of steam-powered warships landed in Japan in 1853, many Japanese who had never seen a ship with a smokestack thought they were seeing dragons. Dragons were part of their belief system, whereas steamships were not.

To counter this tendency toward bias and to make their work more efficient and enable them to give accurate testimony in court, police often receive special training in observation. Likewise, wine and coffee tasters must develop a keen gustatory sense, a process that goes hand in hand with the knowledge they gain about the varieties of wines and coffees available and the factors affecting their flavors, such as soil, climate, and altitude. Musicians hear differently and painters see differently from most of us. These special abilities are developed and enhanced by training, which also affects beliefs, giving one a different set of expectations. Some people are born with such extraordinary abilities and dispositions, which we call talent. As we shall see later, modern neuroscience provides scientific evidence that our beliefs and expectations have a very deep influence on our perceptions.

Eastern contemplative traditions take this even farther. In part 2 we will observe that when our limiting beliefs and expectations are laid aside, we may achieve direct insight into the nature of the mind's participation in all appearances—from rainbows to galaxies.

How Objective Is Science?

If expectations can so dramatically color common situations like these, can they creep in and influence the workings of science, which prides itself on objectivity?

They can and do—and at every level of the scientific enterprise. At one end of the spectrum we tend to think of instruments such as the telescope and microscope as relatively simple devices that provide clear, open windows into worlds hidden from the naked eye. You need only peer into the eyepiece and the moon's craters or a plant's cell structures burst into view. However, both training and theory are necessary for the sophisticated use of these instruments. The researcher needs to know what to expect, what he or she is looking for.

To complicate matters, optical instruments manipulate light and images in ways that both distort the object under study and require special techniques of looking by the operator. Studies in fact show that a small number of trainees simply cannot see properly through a microscope and are therefore discouraged from pursuing fields requiring that skill. (The Dutch naturalist Anton van Leeuwenhoek was one of the developers of the microscope in the early seventeenth century. When human sperm was placed under the lens, he and his colleagues claimed they saw tiny, fully formed humans swimming around!)

What we can see with our instruments—as well as the very instruments and experiments that scientists design—are determined by our expectations of what is out there to be found. Those expectations are based upon beliefs. Just as you believed your date would meet you at a particular time and you mulled over the exciting prospects beforehand, scientists design instruments and experiments based on their beliefs in theories and hypotheses. They bring a lot of extra baggage with them when they step into the laboratory: their various spiritual, cultural, and philosophical backgrounds; their need for funding; their rivalries and disagreements with peers; their ambition; their personal problems; and so on. Many of them are believers in scientific materialism, even if they are unaware of the fact. Their expectations may be unconsciously guided by the tenets of the scientific materialism we examined earlier.

Ideally, the scientist maintains an open mind regarding the process of gathering data and devising and revising theories, but as we shall see, it is virtually impossible to meet such an ideal. The deck has already been stacked. Here is how three distinguished scientists have described the relationship between scientific expectations as expressed in theories and the data that can be collected to corroborate them:

Werner Heisenberg: "What we observe is not nature in itself but nature exposed to our method of questioning."[1]

Niels Bohr: "The procedure of measurement has an essential influence on the conditions on which the very definition of the physical qualities in question rest."[2]

Albert Einstein: "[O]n principle, it is quite wrong to try founding a theory on observable magnitudes alone. In reality the very opposite happens. It is the theory that decides what we can observe."[3]

That seems rather different from the scientific method described previously, "the systematic pursuit of knowledge involving . . . observation and experiment, and the formulation and testing of hypotheses." Furthermore, the implications are significant. Science, especially according to scientific materialism, exists to discover the true nature of a real, independent, freestanding universe out there beyond the reach of our senses. Yet if scientific procedures are based on human beliefs and those beliefs determine our method of questioning—won't that skew the data? How do we know that our "method of questioning" is appropriate? We may like the answers we get simply because they confirm our expectations. We may be asking the wrong questions—questions that reinforce our own biases, obscuring aspects of reality that don't fit our current conceptual pigeonholes.

If we believe in scientific realism—the view that science can describe reality as it exists independently of any observer—it all comes down to the following questions: If there is a single real universe based on fundamental truths, can we know them, and if so, how? And if those truths are accessible to the human mind, are our belief-based theories leading us toward those truths or standing in our way? From its inception, science has taken an optimistic stance, that there *is* truth to be found and we *can* know it. To get there, it has bet on the effectiveness of theorizing and gathering data, or some combination of the two. Throughout the ages scientists and philosophers have examined and debated a wide range of philosophical issues in trying to answer these questions.

DOES THE MIND GET IN THE WAY?

God's role in the universe was not deeply challenged by science until the nineteenth century. By that time the principle of the conservation of energy had become generally accepted: energy can neither be created nor destroyed, merely transformed. This ensured the widespread adoption of physical reductionism and the closure principle. Everything was reduced to the physical, in the form of either matter or energy. Success had given science enormous credibility—it could dance alone. However, the prestige of Newton's classical model, with its foundation in materialism, did not mean it was accepted by all scientists. In fact, it had been questioned all along. The history of science, from ancient Greece up until the nineteenth century, involved not only the devising of theories and collection of data but also philosophical debate. One of the main questions was, How can the universe be best explained—by underlying mechanisms acting behind the scenes (such as mathematics or gravity) or by paying greater attention to what's going on in front of the curtain (theories focusing primarily on empirical, that is, *observed* evidence)?

Going way back, Plato had argued for the existence of an underlying mechanism of universal order that was, for him, a world of pure forms. Aristotle countered, saying that this approach was too abstract, that better answers lay in patterns of data drawn from observation. In the Renaissance, this debate was taken up by Descartes, who favored the formulation of theories (reasoning from the general to the particular, or deduction), and Bacon, who believed one should gather data before theorizing (inductive reasoning). Newton combined these approaches into the concept of the working hypothesis: although he would begin with a theory in mind, he considered it merely hypothetical. If experimental evidence didn't fit, rather than distrusting the evidence he would either change or discard the theory. Like an experienced detective, he kept an open mind, not allowing his intuitive hunches to override the clues and evidence.

Another topic of this debate is found in the meeting of Cardinal Bellarmine with Galileo. The heliocentric model of the planets was, for Galileo, a theory of celestial reality. For him the fact that it produced more accurate data confirmed that the theory depicted the solar system as it was. His inquisitor, Cardinal Bellarmine, was a remarkable man in his own right—no mere demagogue defending the church through blind faith and intimidation alone. He was one of the early articulators of *instrumentalism,* which questions the value of theoretical entities (in this case the heliocentric model), suggesting that theories are only useful for making scientific data intelligible. From that viewpoint, scientific theories are useful because they forge a link between experimental data and human reasoning, but there is no valid way of connecting them to some unseen, underlying reality. There can always be more than one way to skin a cat.

It was Immanuel Kant (1724–1804), perhaps the most important philosopher of his time, who first suggested that human psychology had an impact on scientific theorizing. After examining Newtonian science he concluded that the inherent structure of the human mind, through its concepts and categories, conditioned the formation and interpretation of experience. Since both the theory and the data were filtered through or distorted by the mind, humans had no direct access to things-in-themselves. Science studied not the real universe but only phenomena as packaged by the human senses and mind. The concepts basic to Newtonian science seemed intuitively correct only because the mind's structure produced intuitive notions of time and space.

Buddhist philosophy makes a similar point, but with a twist. As we will see later, the manner in which realities are conceived is related to context—contexts that include the full participation of the mind, the observer. In this view the mind is more than a filter. Although it frames perception through personal and cultural worldviews, it is capable of liberating itself from these limitations through profound insight.

To Kant it appeared that Newton had found the perfect fit between the mind's cognitive structures and experimental data. Theory and data were brought into a cooperative relationship by the mind. Although Kant believed in a transcendental reality—a true world beyond the senses—he concluded that the innate bias of the mind made it unknowable. Our beliefs, represented in mental structures, stand in the way.

However, Kant was ahead of his time. As we saw, during the nineteenth century the Newtonian model was expanded to encompass all areas of science. With such champions as T. H. Huxley and the X Club waving its banner, scientific materialism essentially became the religious creed of the emerging industrialized world. Science and technology were so successful that any philosophical qualms seemed beside the point. Such questions simply fell by the wayside. It was only in the waning years of the nineteenth century, as scientists began to discover flaws in the mechanical view of the universe that the old doubts resurfaced.

DOES REALITY MATTER?

As the limitations of classical, Newtonian science became clear, especially when it came to accounting for atomic and subatomic phenomena, the debate over realism and instrumentalism, enacted three centuries earlier by Galileo and Cardinal Bellarmine, reemerged. Recall the main question: Did successful scientific theories describe an independent reality, or did they simply give the best explanation of observable evidence? If science could make metaphysical statements—say things about reality beyond the physical—it was stepping into the domain of spirituality and myth. In doing so, was science reaching beyond its grasp? Of course, just below the surface lay the more dramatic and complicated questions brought up by Kant: Can such a metaphysical reality, made up of underlying mechanisms, universal laws, things-in-themselves, be understood by the human mind? Or, we might also ask, does such a reality exist at all?

In the eighteenth century, British philosophers called *empiricists*, including John Locke, David Hume, and George Berkeley, took Cardinal Bellarmine's side in the argument. According to Berkeley, "It is not in fact the business of physics or mechanics to establish efficient causes. Mathematical entities have no stable essence in the nature of things; they depend on the notion of the definer."[4] In this view, a connection could not be made between scientific theories, with their laboratory evidence and mathematical formulations, and real, "efficient causes" (universal truths, laws, or mechanisms) beyond the human senses such as gravity, absolute space, or the mathematics of geometry and calculus. Any speculation beyond a theory's empirical evidence was just that, speculation.

Twentieth-century instrumentalists (also called pragmatists, empiricists, and antirealists) developed a variety of critiques aimed at their powerful, well-entrenched realist opponents (also called scientific realists, scientific materialists, and adherents of scientism).[5] The credo of these modern instrumentalists was the same disbelief in metaphysical speculation espoused by Bellarmine, Kant, and the British empiricists. This point of view generally accepts that science is correct in seeking regularities from data and formulating theories based upon those regularities. If, for instance, it can be shown that planets regularly move about the sun in elliptical orbits, that can be accepted as scientific fact. But the reason why they do so—be it Newtonian gravity, curved space-time, or some future theory—is speculation. According to the instrumentalist approach, science should search for the most useful, general, and economical scheme, and stop there, without looking for underlying causes. Although they are a minority, this position has been held by a number of scientists, including the prominent Austrian physicist and philosopher Ernst Mach (1838–1916).

MORE NAGGING DOUBTS

In the 1960s, a new critique of realism suggested that scientists and their theories are strongly influenced by social and cultural paradigms and that scientific revolutions come and go according to shifts in these deeply held worldviews. This implies that science is to some degree dependent on social forces. In other words, is science advancing progressively to discover a single and true underlying physical reality—a grand unified theory (GUT) or theory of everything (TOE)—or merely operating under the influence of cultural and historical patterns? This latter possibility, it was suggested, could be one reason why scientific theories are historically unstable, why new theories are always invalidating their predecessors. In cosmology, for example, the beliefs of the ancient Greeks, based on their cultural worldview, conditioned their conception of the universe. One of their main concerns was the nature and existence of natural laws governing the cosmos and how these might relate to human conduct. The early Greeks sought harmony in the laws governing their city-states, and in cosmology they also saw harmony as the main function of the laws of the universe. Such harmonious laws were central to their culture.

A different set of beliefs in Europe in the Middle Ages placed the Christian God in this key cosmological position. Hierarchy replaced harmony. Then, from the Renaissance until 1900, the beliefs of a gradually industrialized Europe rested on the mechanical view we explored earlier. Now in modern and postmodern society, with all its turbulence, the mechanical view has given way to a host of highly complex, competing theories. This fragmentation in science seems to mirror life today. Therefore social context may indeed produce a strong influence on science.

Another view within the instrumentalist camp is *pluralism*, the belief that the different branches of science with their varied methodologies should have equal status, rather than taking physics to be the cutting edge. According to the current, widely accepted interpretation of science, physics, which regularly speculates on the "real causes" of things, stands at the top. Chemistry then takes its cues from physics, biology from chemistry, the cognitive sciences from biology, and so on. Each field aims to incorporate the truths revealed by physics. For example, current neuroscience generally refuses to consider nonphysical causes for mental phenomena, believing that the mind, if it exists at all, boils down to complex configurations of chemical processes based ultimately on atomic theory (though on a rather outdated interpretation of atomic theory). Pluralism, by contrast, questions that pecking order and its implications. Why cannot neuroscience, for instance, evolve according to its own particular subject matter?

Pluralists also attack the notion, common in society today, that there exists a single, cohesive scientific worldview. In the media we are often given the impression that there is a unified scientific attitude behind such things as governmental decisions on technical issues ranging from global warming to prescription-drug safety. A similar impression is given in the media's treatment of studies and pronouncements issued by university science departments or medical and other research institutes. Such an appearance of general scientific agreement is, according to pluralists and some instrumentalists, either propaganda or wishful thinking based on the hope that some day such a view will be justified.

A more radical position within instrumentalism is held by the modern *pragmatist* school of philosophy. These scholars believe that scientific truth should be based on the usefulness of a theory and nothing else. If knowing the angles by way of Newton's laws enables you to put the eight ball in the side pocket so that you win at pool, then Einstein's theory of curved space-time is irrelevant. Of course, if one's practical aims are more complicated than pool, they may require more sophisticated theories, such as Einstein's curved space-time.

One of the most pointed challenges to the realist position in science questions the meaning of correspondence itself. If scientific theories are models that represent the workings of physical reality, what precisely do we mean by *representation*? In what way is the planetary model of an atom *like* the "real atom"? How does the simple mathematical formula for Newton's famous inverse square law "represent" or correspond to actual gravity? Both the atom and gravity are essentially invisible to us. In what way are such proposed but never seen entities similar to a picture of planets or mathematical symbols on paper? To make such a comparison would be like bragging to someone that my girlfriend looks "exactly like" Marilyn Monroe, only my girlfriend is invisible.

Since these arguments are mainly confined to the halls of academia, they have not greatly threatened scientific realists. For over a century the realist position has been accepted by the majority of scientists as well as the heads of public and private institutions supporting and teaching science. Even so, the more philosophically minded among scientific realists have devised their own sophisticated arguments to parry these attacks of the instrumentalists and other detractors of realism.

CIRCUMSTANTIAL EVIDENCE

Many instrumentalists will accept a certain amount of reasoning beyond the bare data in the creation of scientific theories. (Studies that use a small sample and then apply their findings to a larger population are an example.) In contrast, scientific realism is based on circumstantial evidence, a variety of clues that require reasoning well beyond the data. Taking that step means relying on the creative power of the mind, the imagination. Scientific realists believe that the mind is reasonable, subtle, and wise enough to see beyond patterns to their causes or mechanisms that exist independently of any observer or system of measurement.

From this basic assumption that there must be causes behind perceived regularities, theories are proposed. The validity of this approach, according to philosophers of science who support realism, is based on the strength of the criteria by which theories are evaluated: What determines "strong evidence" for the correctness of a scientific theory? Among these criteria are accurate *prediction* (Does the theory consistently and correctly predict the outcome of experiments?), *coherence* (Does the theory's formulation make sense mathematically, logically, in terms of other scientific norms, and so on?), and *fertility* (Does it lead to new knowledge?). The *practical value* of a theory is another endorsement of its validity (Does it yield new inventions, cures, and so on?). As well, scientists have long valued *simplicity* as a virtue in theories, along with *aesthetic qualities*, such as a theory's "elegance." When scientists propose theories and devise and perform experiments, criteria such as these guide their research.

One prominent philosopher of science, Ernan McMullin, provides the following concise formulation according to four general criteria that support a theory's believability in the realist sense. Since they complement, or go beyond, the purely pragmatic validity of a theory, he calls them "complementary virtues":

Internal virtues: logical consistency, coherence (or "naturalness," absence of *ad hoc* features), causal specificity

External virtues: consonance with other parts of science and (more controversially) consonance with broader worldviews (with metaphysical principles of natural order, for instance)

Diachronic virtues: revealed over time as the theory develops and meets new challenges, fertility, and consilience (unification of scientific domains previously thought to be isolated)

Uniqueness: the absence of credible theoretical alternatives[6]

Obviously these virtues provide the basis for strong circumstantial evidence. Their skillful application would no doubt impress a jury in a criminal case. But do high standards imply or guarantee a correspondence to reality? Suppose someone is on trial for murder, but the body has not been found. Witnesses say they saw the accused attacking the victim, who has not been seen for a month now (external virtues). It appears that the witnesses are sane, that their senses are in working order, and that they are "objective" (internal virtues). The accused has a police record (diachronic virtues). And the police have no credible alternative theory (uniqueness).

But it is still possible, as has indeed happened occasionally in such cases, that the victim will suddenly appear, saying, "Oh, we were just practicing kickboxing down by the river. I went home and the next day flew to the Caribbean for a month's vacation. Only when I returned yesterday did I learn my friend John was accused of killing me!"

Although the victim's surprise appearance completely undermines the prosecution's theory, most of us would consider this case unusual. As members of a jury confronted with such strong circumstantial evidence, we would probably have found the accused guilty. Realism backed by this type of reasoning has been generally accepted in Western society for centuries. As we saw, it became progressively stronger as science took over the role of truth-definer from Christianity, at least in the minds of those with faith in the scientific community and its methods of inquiry. Many of us are now thoroughly saturated with this approach to understanding—it is part of what we call "common sense," a collection of shared assumptions we accept without question. Instrumentalists such as Cardinal Bellarmine had an easier time opposing scientific realism because they could always fall back on the "will of God" to explain the "why" for what is invisible to our senses. But based on an antirealist position, how strong a challenge can one mount against accepted scientific theories based on the belief in realism?

TOO MANY CIRCUMSTANCES?

Let's look at electricity. What is it? What do we know about it? We can see it (lightning, sparks from a car battery), touch it (the experience of electrical shock), hear it (humming, crackling), and maybe smell it (the ozone created by lightning). By placing our tongue across the terminals of a battery we might even taste it. Furthermore, we know that it causes things to happen. Thus, something we commonly call electricity can be perceived by our senses and we can describe it in words. We could stop right there, but scientific realism tells us there must be a lot more to it than that.

An idea among scientists that electricity was some kind of "flow" dates from the early eighteenth century, and up until the nineteenth century scientists believed electricity was a fluid. In 1753, the French scientist Charles Dufay said that electricity consisted of two fluids, one negative and one positive. Benjamin Franklin, who in 1752 discovered that lightning was an electrical conductor, believed electricity was a single fluid with negative and positive states. In 1785, Charles-Augustin de Coulumb applied Newton's inverse square law to magnetic and electrical forces, showing that, like gravity, electrical force varied with distance according to a mathematical formula. In 1827, Georg Simon Ohm described electricity in terms of three characteristics: force, current, and resistance. In 1831, Michael Faraday theorized that electricity (as well as magnetism) operates as forces within a field. In 1857, Gustav Kirchoff discovered electromagnetic signals can travel at the speed of light. And in 1898, J. J. Thomson discovered the electron.

Today, if you take a class in basic electronics or an undergraduate course in physics, electricity will be described to you as essentially a flow of electrons. It will be explained that conductors such as copper wire are materials made up of atoms that have many open slots available in their outer shells. These are the outer reaches of each atom's minuscule "solar system," where some of its electrons can shift from atom to atom. When a wire conductor is hooked up to a battery, these electrons begin moving from atom to atom, creating an electric current. (Notice that the atomic model used in this explanation is not the most recent one from our list in the previous chapter.)

Thus, scientists have been looking for "electricity" for a long time, finding a variety of answers to the question What is it? All along they have been using this mysterious substance to create marvelous gadgets such as the lightning rod (1750s), telegraph (1840s), incandescent lamp

(1870s), telephone (1876), radio (1895), television (1930s), and computers (1940s). If today we say electricity is a flow of electrons, what, then, is an electron?[7] Is the electron something scientists "discovered" or was it something they were *looking for* to explain a variety of seemingly related phenomena?

Using the model, picture, or framework currently accepted, an electron is an extremely small, subatomic particle with a very low mass, which "circles" the nucleus of an atom and is classified as a lepton. Since it cannot be seen—we cannot magnify a single atom to confirm that it looks like a little solar system—the electron's existence has been determined by inference. It is deemed to exist because of (1) the way things seem to work in laboratory experiments (that often lead to practical applications), and (2) the way these data either suggest theories or conform to previously conceived theories. It is pinned down by reasoning, by the criteria that support the validity of scientific theory from the realist perspective.

However, it is important to note that as understood within current theories of physics, the electron exists in context with the atom and its constituents. At present, it is thought to have mass and charge but no spatial dimensions and no definite location and momentum at any precise moment in time. The theories in use when the telegraph, light bulb, and telephone were invented made no mention of the electron. So there is no guarantee that future theories of electricity will do so either. Yet presumably light bulbs will still glow.

How does electricity and its electron stack up in light of McMullin's complementary virtues? As to internal virtues (logical consistency, coherence, and so on), we can assume that these theories were well reasoned. Up until the twentieth century, concepts such as electrical charges and electromagnetic fields were elaborated with great care by the likes of James Clerk Maxwell and Hermann Helmholtz. Peer review, especially with such eminent peers, places a relatively strong guarantee on the coherence of accepted scientific theories. However, the picture is less consistent regarding external virtues. As we've seen, the view of electricity changed frequently prior to the twentieth century, and dramatically thereafter with quantum physics and the standard atomic model. In terms of diachronic virtues, the early theory of electricity did evolve to meet new challenges, such as the gradual development of the notion of electrical charge and field theory. However, since 1900 it is hard to tell where the atomic theories underlying the electron have left us.

As for uniqueness, the terms *electricity* and *electron* have been applied to a wide range of theoretical entities, that is, to the kind of entities proposed by scientists that cannot be observed directly. Today there is no unique theory in physics accounting for electricity. To be fair, criteria such as the complementary virtues are best applied to a single theory rather than a string of them. However, since these virtues supposedly support the reality of some underlying cause or mechanism (here, the electron), even so crude an examination as this is revealing.

The concept of the electron, embedded within current atomic theory, may give a more useful picture of electricity than earlier theories. Yet is there any guarantee that the electron is anything more than a convenient model around which to explain the characteristics of electricity discovered so far? Furthermore, can we be certain that a more practical model won't be discovered in the future, making the electron obsolete? Are we coming closer and closer with each new theory to the "true understanding" of the "real electron," or are new scientific worldviews, perhaps based on new cultural paradigms, simply wiping out old theories and replacing them with new ones?

How Do We Choose a Reality?

What happens when we have more than one solid theory to explain a single phenomenon? In the case of gravity, for example, does scientific evidence force us to choose between gravity as a force of attraction (Newton's view) and gravity as a consequence of the shape of space-time (Einstein's idea)? The answer is no. In most circumstances both of these theories adequately account for the effects we call gravity. Situations are called *underdetermined* when the evidence doesn't compel us to accept one from among competing scientific theories. Such situations are surprisingly common in science.

Recall Einstein's Nobel Prize–winning theory of the photoelectric effect, which demonstrated how photons were emitted according to quantum principles. In 1969 a team of scientists proposed an entirely different explanation of the photoelectric effect. Their theory, which uses both classical wave theory and quantum physics, accounts for the evidence just as well as Einstein's did. In addition, several other classical and semiclassical solutions have been found that explain specific sub-atomic events just as well as quantum physics. How are we to choose among such competing theories?

The fact of underdetermination has some disquieting implications for science. As scientists choose among competing theories (or make alterations so that old theories better suit new data), they confront the problem of circularity: The "objective criteria" they use in making such choices are really chosen on the basis of their subjective, personal views, their beliefs. If they lean in an instrumentalist direction, one set of criteria will be used. If they lean toward realism, others will be chosen. It was Einstein's belief in the realist tenets of classical physics such as locality that led him to adopt the EPR position. There he assumed that only local effects could be in operation, whereas it was later shown that this assumption failed to account for the evidence and that nonlocality (entanglement) succeeded. It never occurred to Einstein to question his belief in locality or to test it in some way. At rock bottom Albert Einstein was a scientific realist. In the EPR paradox, his unexamined beliefs came around full circle, blinding him to the error in his hypothesis.

CHOICE OR NO CHOICE?

When confronted with competing theories, or when altering components to give the theory wider scope, scientists typically base their choices on considerations such as the complementary virtues. That choice itself is a restricted one, because it is based on the question Which of the competing theories or versions of a theory best corresponds to "reality"? Criteria such as simplicity, elegance, coherence, fertility, and so forth have been consolidated over time as the means of judging theories specifically in terms of their correspondence to reality. Scientific realists who seek some combination of these criteria do so because they believe that true theories have some combination of these qualities. An instrumentalist, in contrast, would be more interested in the practical values of a proposed theory: is the concept "electron" useful for describing electricity?

The debate between realists and instrumentalists in science leaves no clear winner. We are seemingly forced to choose between absolutism (things are really real) and nihilism (nothing is real). The circularity that underlies the realist position implies that scientific realism may go too far in its extrapolations beyond scientific data. No matter how objective or unbiased the experimental procedure, the underlying assumptions are not objective, and they lead to experiments and technology that may merely justify the initial beliefs. Believing shapes seeing. Both a theory and its instruments and procedures are essentially a means of selecting certain data from among the mass of information available.

This is not unlike the so-called glass ceiling used in hiring practices. If I, as an employer, assume that men make better executives than women, out of the homogeneous population of males and females I will select mostly men. I won't be interested in evidence showing that women make good executives. That will be filtered out from the start. If I then lose business to competitors with women in executive positions, my belief will prove to be unjustified. In like manner, we have seen ample evidence that realist theories in science have made unjustified claims. (Even the claim that these failures are part of a process of elimination on a progressive road to truth is not necessarily justified by the record.)

While the claims of realism are unjustified, those of antirealism leave us impotent. Such methodical skepticism keeps us in ignorance about many things we wish to know scientifically and doesn't help us much in our more philosophical exploration of the true limits of our understanding. Disbelief becomes a form of blindness. In defense of antirealism, the empiricist philosopher Bas van Fraassen said, "Science allows perfectly well for the sceptical [sic] discipline that accepts the appearances alone as real, and all the rest as a unifying myth to light our path."[8] That is, experimental evidence alone is the heart of science, but "all the rest"—speculation beyond the evidence—is only useful to prod science forward. Unfortunately, between the nihilism of the skeptics and the absolutist position of realists we are left in the lurch. The realist must always cry out in frustration, "Our path to *where*?" while the antirealist whistles nervously in the dark.

Perhaps this discussion allows us to be a bit more objective about science itself. Modern education, the news media, and conventional wisdom have combined to convince us that science is objective in the sense of being value-free—a trustworthy tiller to steer us through the choppy waters of ignorance, prejudice, and superstition. Yet as we've seen, science has its own history, one that sometimes burdens it with prejudice. Philosophers have presented challenges to scientific thinking, but has science responded to them fully? What is the proper balance between data and theory, induction and deduction? If we favor a particular theory, will we filter out important evidence pointing to other, more fruitful directions?

What of the challenges of instrumentalists, empiricists, and pragmatists, who insist that science explore the utility of theories and not be tempted to read too much into them in terms of natural laws and universal truths? We also observed how underdetermination—cases where more than one theory successfully accounts for experimental data—throws into question the whole proposition that there exist unique, underlying mechanisms operating behind observed phenomena. Have scientific realists—who rely on a theory's predictive ability, coherence, fertility, and so forth—made a convincing case for the existence of such universal truths? Or are we, as Immanuel Kant suggests, left with no choice at all except to follow the dictates of the human psyche, a built-in human mentality that casts its reflections on phenomena, leaving an indelible human imprint?

For the most part, science sweeps such doubts under the rug and moves vigorously forward. However, a closer examination of the central figure in all of this scientific thinking, the mind itself, may suggest a solution. Could it be that the mind has an intimate connection with physical phenomena that could clarify not only the nature of the scientific quest but the universe science seeks to understand?

The Mind behind the Eyepiece
Limits on Scientific Objectivity 4

THERE IS in science a term called *black box*. One dictionary definition of a black box is "Anything having a complex function that can be observed but whose inner workings are mysterious or unknown."[1] If you have no knowledge of electronics, then your CD player, stereo amplifier, television, and even your telephone are all black boxes. You know only that some kind of signal, say from your TV antenna or cable, enters the box at one end, then information usable by your mind and senses exits the other: a newscast, sports program, film, or sitcom. Just how this happens is a mystery.

Oddly enough, the very science that created all of these marvelous boxes has a mega–black box of its own: the mind. To make matters worse, science has never quite been able to believe in this black box. Generally it has either been suspicious of internal, subjective experience, denied its very existence, or avoided the subject altogether—mostly the latter. There is obviously some irony here. What scientific idea has ever come from anywhere else but the mind of a scientist? Likewise, what observation in the laboratory has been made by any other faculty than the mind? It doesn't take deep thinking to reach the conclusion that the most important, indeed the *fundamental* scientific instrument is and always has been the human mind. That is where theories originate, experiments are designed, data are observed—and it is the mind that comes to scientific conclusions: "This theory works," or "That theory stinks!" So since scientists obviously need intimate knowledge of the instruments they design to aid in their explorations—telescopes, microscopes, particle accelerators, and all the rest—how is it possible that they have so long postponed a thorough examination of their own minds?

We have already seen some of the reasons why. Recall that science as it emerged in the Renaissance sought truth "out there" in the heavens, the objective universe. Natural philosophers were seeking to know the mind of God by way of his creation. Their reasoning was simple: understand the mind of the Great Mechanic by understanding the nature of the great machine, the objective world of nature that he created and maintains. The opposite, "inner" direction, on the other hand, was toward the center of human identity, where many feared the Devil was lurking, ready to possess people's souls. This scheme of things, so pervasive in the Renaissance mind, provided the basis for objectivism. Just by making knowledge of an independent, objective reality outside of our minds the goal of science, the subjective side of nature was automatically shunned.

Recall also that the Renaissance was an era of witch-hunting, the time of the Inquisition and the rise of Protestantism. The mind was indeed considered dangerous territory, susceptible to the temptations of the Devil. In a Europe dominated by such fearful beliefs, a science of the mind was out of the question. One natural philosopher, René Descartes, did at least speculate on the nature of the mind. Descartes believed that the soul could act voluntarily on the body through the pineal gland, located in the central part of the brain. All the other actions of the body were mere reflexes. This dualist view—with a nonmaterial mind or soul acting independently on matter (the body)—had strong influence on science for centuries to come. However, it also played an important part in delaying the appearance of scientific investigation of the mind, given that half of Descartes's formula, the soul, was a metaphysical issue considered off-limits to science.

Ironically, Christianity itself had all along been exploring the mind. Early in its history, contemplative traditions developed within the Catholic Church. This went on largely behind the scenes in monasteries and hermitages. Turning inward, Christian mystics sought intimate contact with the divine, even to merge with the mind of God. The extensive literature of Christian mysticism suggests an analysis of the mind based on direct experience, on introspection. Unfortunately, this contemplative tradition was largely outside the religious mainstream. The number of monastics actually engaged in contemplation was small. Nevertheless, they did exert an influence on the church as a whole, and some of the leading figures were well known. For example, both Saint Teresa of Ávila (1515–82) and Saint John of the Cross (1542–91) founded monastic orders in Spain.

When science finally cast its glance inward to the mind, an examination of these exceptional states of consciousness might have been fruitful. However, even contemporary mind science has for the most part studied only normal and subnormal states of consciousness, and invariably viewed them from the perspective of an outside observer.

MIND OR BRAIN?

It wasn't until the late nineteenth century that science attempted a formal study of the mind. Given the enormous influence of scientific materialism, it is not surprising that a physical approach—the study of behavior and the brain, the "gray matter"—held sway. By the early twentieth century, nonmaterial qualities attributed to the mind (thoughts, feelings, images, dreams, and so on) were neatly avoided by correlating them to the physical brain, with its internal physiology, and to physical behavior. Thus, mind was simply redefined as the brain.

The influence of this approach can be seen in the early activities of the two foremost psychological theorists of that time, Sigmund Freud and William James. Freud began his career with the intention of creating a theory of the psyche based on materialist principles, on physiology. He was a teacher of neuropathology, and his early research included a significant study of the brain's medulla. Later, however, clinical experience with patients suffering from hysteria suggested to him that the mind, rather than the brain, might provide a more fertile ground for his research. From that point on he shifted from the study of brain function and structure to the investigation of first-person accounts of the experiences of individual patients. His major contribution to the field of psychiatry—psychoanalysis, which is popularly associated with the patient on the couch relating his experiences to the seated psychoanalyst—was based on the subjective experiences of his patients. Rather than dissect brain tissue, he listened carefully to what his patients said about their minds.

The American psychologist William James (1842–1910) also began in the physical sciences—chemistry, biology, and medicine, including studies in Germany with the renowned physicist Hermann Helmholtz. James, too, intended to create a science of the mind informed by physiology, and is credited with founding the first laboratory for experimental psychology in the United States. In his early writings he emphasized the manner in which physical sensations influenced the mind, viewing the mind as a product of biology. However, James moved on to a vision of the mind that was far more embracing than that. Eventually defining himself as a pragmatic empiricist, he came to the conclusion that science should always concern itself with direct experience, including the first-person experience of the mind. In other words, personal experience regarding mental phenomena was crucial.

This assertion signaled a major expansion of the subject matter usually studied by empiricists. It led to a view that, like Freud's, accepted the importance of first-person accounts of subjective, mental experience. But James also believed that introspection, the systematic exploration of one's own inner experience, could become the foremost tool for understanding the mind. (Here, his view parallels the contemplative approaches we will look at in part 2. Referring to effects of spiritual practice on the individual, James said, "Knowledge we could never attain, remaining what we are, may be attainable in consequences of higher powers and a higher life, which we may morally achieve.")[2]

James's belief amounted to reestablishing the presence of mental phenomena in the natural world. For him, the natural world—and recall that early on scientists were called "natural philosophers"—should include both the objective, physical world and the subjective, inner realm of the mind. Making his view of mind science thoroughly holistic, James also accepted comparative research with animal behavior and experimental brain science as complementary to introspection, which he considered fundamental.

James even went so far as to apply his empirical approach to the study of spirituality. Abandoning philosophical theology, he sought to understand "the immediate content of religious consciousness." After studying the experiences of contemplatives from diverse traditions, James concluded,

> Our normal waking consciousness, rational consciousness as we call it, is but one special type of consciousness, whilst all about it, parted from it by the filmiest of screens, there lie potential forms of consciousness entirely different. . . . No account of the universe in its totality can be final which leaves these other forms of consciousness quite disregarded.[3]

Historically speaking, however, William James was a flash of lightning in the dark of night. He was influential and is still considered the foremost pioneer of American psychology. However, only those of his views compatible with scientific materialism aroused enthusiasm among his successors. In the late nineteenth and early twentieth centuries a fledgling introspective school of psychology did emerge, but it was burdened by very restricted experimental approaches. Many of the tests conducted on human subjects, which were designed to obtain objective knowledge of subjective states, were extremely simple and tedious, and James dissociated himself from this approach. Unfortunately, although James was somewhat acquainted with the introspective methods used in the practices of Hindus and Buddhists, he was unaware of the high degree of attentional stability and introspective depth they claimed to achieve. Therefore, he concluded that attention could only be sustained at the highest level for a few seconds, not enough to successfully fathom the nature of consciousness. This conclusion is still accepted by science to the present day.

MERE WASTE . . . NOTHING AT ALL

The way we pay attention to things plays a crucial role in how we view reality—how we decide what exists and what doesn't. If our attention is glued to a television program, perhaps a hotly contested sporting event, a dog barking outside our door may go completely unnoticed. For that moment, as far as we are concerned, the dog is nonexistent. Should that dog enter our home and bite our ankle, at that moment the TV program would fade from our personal world and the dog would become very real. "Our belief and attention are the same fact," wrote William James. "For the moment, what we attend to is reality . . ." In contrast, those things not attended to "are treated as if they were mere waste, equivalent to nothing at all."[4] Sad to say, this is precisely what happened to James's all-embracing approach to the mind and to the fledgling introspectionist school in the early twentieth century. Fascinated with the success of physical science, psychological theorists placed their attention on physical approaches to the mind, refusing to believe anything could be gained by studying consciousness from the inside—out of sight, out of mind, out of existence. Scientists of the mind soon shifted their attention to something they could measure: behavior.

SALIVATING CANINES

The founder of behaviorism, the American psychologist John B. Watson, wrote in 1924: "Behaviorism claims that 'consciousness' is neither a definable nor a usable concept; that it is merely another word for the 'soul' of more ancient times. The old psychology is thus dominated by a subtle kind of religious philosophy."[5] For Watson, consciousness was a modern substitute for the "soul," a spiritual notion and therefore not acceptable to a science of the mind. Valid results could only be based upon observable, measurable data. Behaviorism, which dominated research on the mind from World War I until well into the 1950s, is based upon the study of stimulus and response. Its most famous model is "Pavlov's dog," trained by the Russian physiologist Ivan Pavlov to salivate at the sound of a bell. To the behaviorists, conditioned reflexes of this sort were the key to understanding the mind. Their model placed emphasis on mechanisms of behavior by which the human organism responded to the outer environment. The mind that thought, emoted, felt, created images, and so on was viewed as a set of conditioned, mechanical responses. Since these could be measured with some consistency by outside, objective observers, and because such measurements more or less corresponded to the methods of physicists and chemists, the approach was assumed to be the correct one. The complexities and inconsistencies of psychoanalysis and introspection were neatly done away with by discrediting their "unobjective methodologies." The behaviorists didn't even regard the mind as a black box, they declared that the box of the mind—the realm of subjective experience—didn't even exist, at least from a scientific perspective.

According to this approach, psychological therapies would be replaced by behavioral therapy—behavior modification—which uses systems of rewards and punishments (negative and positive reinforcement) to "improve" behavior. In Stanley Kubric's famous film, *A Clockwork Orange* (based on the novel by Anthony Burgess), a habitual criminal is reconditioned by drug therapy and repeated viewing of filmed scenes of violence in order to wean him from his violent habits. His reflexes are reprogrammed. However, the therapy only works superficially and temporarily because his mind is much deeper and more complex than mere reflexes. Soon he is back to his old tricks. Likewise, if traditional definitions of the mind were too hazy and mysterious, this new definition as behavior mechanisms proved to be an oversimplification. By the 1960s it was becoming clear that purposeful behavior, with its complex choices and motivations, played too important a part in human mental life to be excluded by the beliefs underlying behaviorism. This realization ushered in cognitive psychology.

THIS DOES NOT COMPUTE ...

Because human mental life was more than mere stimulus and response, a more sophisticated model was needed to describe it. That new model was drawn from the fields of information processing and computer technology. Information processing was the brainchild of the American logician, mathematician, and physicist Charles S. Peirce (1839–1914). Peirce developed a description of the way information is processed in terms of its syntactic (structural), semantic (meaningful), and pragmatic functions. This systematic examination of information led to constructs such as those found in modern computers, whose basic modules are a processor, memory, receptor, and effector. Flowcharts described how information moved among these modules.

In cognitive psychology such models are believed to imitate the workings of the human mind. Cognitive activity is viewed in the abstract as forms of mental representation—symbols, images, ideas, and patterns—all stripped of subjective meaning and having no connection to consciousness. Here people are likened to the mechanical robots of early science fiction movies that, when confronted by human responses, would bleat in a monotone, "This does not compute . . . this does not compute." Note that reducing the mind to a computer model is but one more example of physical reductionism. And reducing it to mere gadgetry had been done before: at the beginning of the Scientific Revolution the mind was compared to a hydraulic system, and in the early twentieth century to a telephone switchboard.

Over the past half century, cognitive psychology has joined with neurobiology, which studies the physical structure and interactions of the brain and nervous system, to form contemporary brain science. Here, cognitive psychology attempts to bridge the gap between the purely physical and the varied functions of the brain such as memory, visualization, perception, and even consciousness. Since the mid-nineteenth century, correlations or correspondences between physical brain structure and mental function have been observed and catalogued. Much of this data has come from observing mental impairments resulting from injuries to the brain. For instance, it was noted early on that speech defects ensued after injuries to certain regions of the cortex. Researchers then theorized that there existed one-to-one relationships between activity in a specific region of the brain and specific mental functioning. Later, it became clear that the picture was much more complex: interconnectivity between various regions of the brain apparently acts in an overlapping, cooperative manner to produce the physical correlates to mental phenomena.

Over time, increasingly sophisticated instruments were designed to more closely observe brain functions and map these correlations between brain structures and functions. Today, for example, functional magnetic resonance imaging (fMRI) is to some degree able to show brain activity happening in real time. The researcher can view images of the brain in action on a computer screen. Such technology has enabled scientists to map significant areas of the cortex. This is a field still in its infancy, and a complete and accurate list of correlations between brain functions and mental states is a long way off.

The most ambitious researchers, including the late Francis Crick (who won the Nobel Prize for his contribution to the discovery of the double-helical structure of DNA), believe that brain science based on this kind of research will show that all mental qualities popularly conceived of as nonphysical—emotions, thoughts, and consciousness itself—are properties that emerge from the functioning of physical brain structures. According to this belief, correlates between behavior and the brain's observed physical functioning do, or will soon, provide proof of this hypothesis. It would seem to make sense: stimulate neurons in some part of the brain and subjects experience some alteration in their field of awareness or in their capacity to function. This is not essentially different from the changes of consciousness brought on by a couple of stiff drinks of alcohol—matter affects mind. According to this view, to be drunk is merely to have your neurons tweaked by a substance with the chemical symbol CH_3CH_2OH. Let's look a little deeper.

YOU ARE YOUR NEURONS?

Of course, this line of reasoning can be taken a lot farther. The "you" who are drunk—comprising all the functions of the personality such as your values, opinions, habits, and so forth—can be viewed solely as a product of the structure and functioning of the brain and the nervous system. The very conscious awareness that allows us to *know* anything may turn out to be, as one theory suggests, a global unity imposed continuously on neurons in a number of areas of the brain. Imagining those neurons to be like sets of lights draped upon a Christmas tree: as long as some of the light groups are glowing, there is consciousness. Indeed, according to this hypothesis such continuous firing of neurons *is* consciousness.

For those driven to find exclusively physical explanations—"hard facts" to solve all riddles—this approach is appealing. Yet clearly it is but a modern version of a very old idea: "No phosphorus, no thought." Cognitive psychology today is part of the historical thrust of scientific materialism with its fixation on the physical. Based on that belief, it was inevitable that brain correlates would be used to redefine subjective qualities and consciousness. It is revealing, though, to observe just how brain scientists have applied this belief historically: In the early days of neuroscience it was declared that mental phenomena differing from physical brain structures were nonexistent for the very reason that they were identical with brain states. Because physical correlates had been discovered and mental phenomena could be reduced to them, there was no more need to believe these nonphysical qualities existed as something separate. Later, when the picture of one-to-one relationships between brain function and mental phenomena was clouded by new advances in neuroscience, the opposite tack was taken: separate mental phenomena were nonexistent because they were *not identical* with brain states. In either case, only brain states were real. The common thread in both approaches is nothing more than the belief that only the physical exists. However, does accepting that some sort of correlation exists between physical brain structure and conscious states necessarily imply that consciousness and subjective states are only physical?

THE SAME COIN?

We have already seen examples of the difficulties that ensue when scientists dogmatically adhere to unexamined beliefs. Historically, these prejudices are some of the greatest impediments to scientific research. If for a moment we take a skeptical stance toward the beliefs of scientific materialism, it quickly becomes clear that the existence of correlations between brain structures and subjective mental states does not prove the two are identical. (The difficulty in bridging this gap is called the "hard problem" of neural correlates of consciousness, a term introduced in 1995 by the philosopher David J. Chalmers.) Back when William James studied this problem, he proposed three possibilities:

1. The brain produces thoughts, as an electric current produces light.
2. The brain releases, or permits, mental events, as the trigger of a crossbow releases an arrow by removing the obstacle that holds the string.
3. The brain transmits thoughts, as light hits a prism, thereby transmitting a surprising spectrum of colors.[6]

Research in neuroscience thus far is compatible with all three of these hypotheses, yet the neuroscientific community has considered only the first, the one conforming to the tenets of scientific materialism. (It is worth noting that Augustine once proposed four alternatives, all compatible with biblical scripture, for the origin of the soul.[7] But, to his credit, he declared, "It is fitting that no one of the four be affirmed without good reason.") By opting only for the first alternative, brain scientists have merely defined away the problem. To even begin to provide a meaningful bridge over the gap between brain structures and subjective mental experience, a more subtle understanding of these subjective states must be obtained, even if they are considered to be illusory. Given the rigors of the scientific enterprise, it isn't very convincing to declare that some phenomenon is like "two sides of the same coin" if you have examined only one side.

By assuming that mind equals brain—that all subjective states boil down to physical functioning—brain science has placed itself in a difficult position. If mental phenomena are emergent or secondary properties of brain tissue, their relation to the brain is fundamentally different from every other emergent property and function observed in nature. In the case of the properties of water (liquidity and solidity), or in the case of photosynthesis in plants, or the digestive process in animals, the emergent properties respectively of H_2O, plant physiology, and animal digestion, can be observed directly. Ice, green leaves, and digestive products can be seen. How could such processes be detected otherwise? But in the theory of brain correlates to mental states, such observation is missing: no mental states are observed when examining brain tissue, and conversely the brain is not observed as part of the experience of a mental state. The idea that mental processes and their correlated brain functions are equivalent is a hypothesis that is simply not validated, yet it is one that scientific materialists find appealing and therefore often accept without question.

However, the most difficult problem with the theory of brain correlates is how exactly the brain produces conscious states. Note that until recently the terms *consciousness* and *awareness* did not even appear in most textbooks of the cognitive sciences. Despite all of the marvelous research done on the material side of the question, cognitive science has avoided the rigorous, introspective observation of the subjective states supposedly correlated to brain processes and structure. This situation would be tantamount to zoologists theorizing that cellular organisms grow by cell division (cell mitosis) but having no interest in, and therefore no means of observing, cells themselves. Of course, zoologists do have such interests and therefore have participated in the development of the microscope.

Indeed, all important advances in science have been made through new instruments, yet brain science has no technological means of detecting consciousness or any kind of subjective mental experience. Nor does it have any means of demonstrating how the brain produces consciousness. By relying on the argument of mere correlations between mental phenomena and brain physiology, cognitive psychologists remind us of astrologers, who rely on correlates between patterns in the heavens and events on earth, rather than astronomers, who have actually explored the skies scientifically with telescopes.

Even proposals for a "psychometer," an instrument to detect consciousness, are riddled with problems introduced by the built-in prejudices of scientific materialism regarding consciousness. What would such an instrument directed at physical evidence detect? By picking up electrochemical signals, temperature changes, and so forth, it would again only be detecting physical data. Furthermore, if such a device could also somehow show a real connection with a subjective mental state, what new information could it provide? Would it be able to distinguish between a brain state as the *cause* of a subjective experience and a brain state as being the *phenomenon itself*? The interpretation would seem to fall upon the subjective opinion of the researcher, not on objective evidence.

FREEING THE SCIENTIFIC MIND

Clearly, any research into the nature of consciousness must begin with at least a tentative admission, a working hypothesis, that subjective mental states exist. Whatever consciousness may be, without it all of us, including brain scientists, would either be in dreamless sleep or a coma. It follows that any method used to explore the mind-brain nexus must incorporate subjective reports from individuals. Seen in this light, cognitive science has all along been approaching its subject matter from a paradoxical perspective, relying on first-person observations by the scientist, but not explicitly incorporating them into the scientific method. If it claims to demonstrate physical correlations to mental phenomena and consciousness, such research should begin with a rigorous, firsthand investigation of consciousness itself.

Logically this could be done via a combination of first-person reports—such as those conducted in psychological testing interviews and psychotherapy—and introspection, ideally by the researchers themselves. Shouldn't cognitive scientists first be experts on their own consciousness, deeply exploring their subjective nature, before they tackle the complexities of the mind-brain connection? Given the rigors of science, wouldn't such self-knowledge be useful for scientists in general? After all, the scientific mind behind the eyepiece of a physical instrument (and behind the devising of theories) is the fundamental instrument of all science. Must not this ultimate black box be opened and carefully examined if science wants to be certain that its theories and data are something more than complex imaginings or projections?

In fact, recent research in neuroscience strongly suggests that there is an intimate connection between imagination and perception, that the two functions share physical structures and patterns within the brain (operating in a very rapid and complex fashion). There is a wide consensus among neuroscientists that the ability to produce and manipulate images is derived from the very same neural capacities as those involved in high-level vision and cognition. The implication—reminiscent of the possibility that believing can be seeing—is that conceptual superimpositions saturate our normal, perceptual experience of the world. According to one view:

> Perception is demonstrably constrained and shaped by the concurrent higher cognitive memories, expectations, and preparations for action. For us, here, this means that what is endogenous (self-activated memories and predispositions, for example), and hence the manifestation of the imaginary dimensions, is always part of perception. Conversely the generation of imagery is not a different, or separate, stream but constitutive of the normal flow of life.[8]

As an example, such internal, mental influences, which can bias our perception, are often experienced in our perceptions of time. We say "a watched pot never boils" in reference to the way time drags on when we have a strong desire that it pass quickly or when we simply concentrate closely on time's passage. Conversely, we say "time flies when you're having fun" to express the opposite perception of time when our minds are directed toward pleasurable activities. Is time, which is so important to everything we do, uniform or not? Human beings, using their minds, invented the clock to make time uniform. The artist Salvador Dalí painted liquid clocks that dripped and drooped. Perhaps he was suggesting that humans could just as easily make clocks that produce quirky, nonlinear time, and that our imaginations are the key to our perceptions.

An intimate knowledge of our mental black boxes, our minds, is vital if science is to make genuine headway in understanding the relationship between the mind and scientific exploration. In that regard it should be clear that the human mind is so pervasive, creative, and all-encompassing as to make attempts to bottle it in the small vial of scientific materialism seem ridiculous. If science cannot emerge from this bottled-up view of the mind, it may never be able to draw meaningful conclusions about the crucial relationship between the mind and the brain.

The preceding discussion should make it clear that science's attitude toward the mind has been hampered by historical baggage. According to the dictates of its Christian background, science explored outer, objective phenomena and avoided the inner, subjective realm. Lack of self knowledge hampered scientists by blinding them to subjective distortions that have prejudiced the scientific enterprise. For example, is there really a strict dividing line between the "inner" and "outer"? Or is this view merely the product of an ingrained mental habit?

Both James and Freud acknowledged the importance of subjective experience for understanding the mind, without denying the complementary data provided by physiological studies. However, such openmindedness was swept aside by behaviorism, and this trend continues today in the cognitive sciences. Simple mechanical models have been replaced by the flow charts of information processing. But the suppositions of cognitive science and its equation of mental states with brain activity and behavior have not been tested according to the rigors required in other scientific fields. Reticence about exploring the mind through introspection keeps cognitive scientists largely in ignorance of the mental states the brain is supposedly producing on its own. Thus, a complementary approach—comprising first-person contemplative explorations of the mind, interviews with the contemplative subject, and physiological data on the activity of the brain and nervous system—holds promise for scientific exploration of the mind as well as for an eventual reappraisal of scientific methodology in general.

Masquerade 5
Scientific Materialism Poses as Science

U P UNTIL NOW we have discussed scientific materialism mainly as a philosophy—one strongly influencing our commonsense notions of things—that developed gradually and somewhat covertly beginning in the Renaissance. Given its importance for all of us, it is a pity that scientific materialism cannot be given some less ponderous, more evocative name. Suggestive titles such as "technocracy" and "Scientology" have already been taken. The latter would have fit well if interpreted as a combination of "science" and "theology," because, as we shall see in more detail, scientific materialism has strong theological overtones.

Scientific materialism cannot be found in common dictionaries nor as a heading in standard encyclopedias. That is one reason why this powerful philosophy, which has dominated scientific thinking for over a century and a half, is practically unknown to the public. However, the main reason is that scientific materialism is usually confused with science itself. By presenting itself as science, scientific materialism acquires the prestige of science while avoiding scrutiny into those areas where they differ. The influence of scientific materialism is therefore both pervasive (like science) and subtle to the extent that we are unaware of how much we are under its spell.

We have seen how the beliefs of scientific materialism developed historically and how many of them came to grief in the early twentieth century, outmoded and undermined by the discoveries of modern physics. Clearly, the principles of scientific materialism were not arrived at scientifically—through "the recognition and formulation of a problem, the collection of data through observation and experiment, and the formulation and testing of hypotheses."[1] They merged with science as part of its cultural and philosophical context beginning in the Renaissance. They were baggage from ancient Greek atomism, scientific views drawn from Christianity, and nineteenth-century European materialism. For a while they were important guiding principles of science, supporting the early exploration of the physical side of nature. Had the use of scientific materialism ended there, as a basis for hypotheses regarding physical nature, science would have been well served. Unfortunately, men such as T. H. Huxley were tempted to go beyond the scientific evidence to create a metaphysical dogma, a "church scientific."

THE CHURCH OF SCIENCE

Let's compare our definition of the scientific method with that of the word *dogma*. From its Latin roots, *dogma* means "an opinion, that which one believes in." In theology it is defined as "a doctrine or body of doctrines formally and authoritatively affirmed." The term *dogmatism* is defined as "[A]ssertion of opinion, usually without reference to evidence."[2] In terms of its functioning, a dogma is a coherent, universally applied worldview consisting of a collection of beliefs and attitudes that call for a person's intellectual and emotional allegiance. Dogma, therefore, has a power over individuals and communities far greater than the influence of mere facts and fact-related theories. Indeed, a dogma may prevail despite the most obvious contrary evidence, and commitment to a dogma may grow all the more zealous when obstacles are met. Therefore dogmatists often appear incapable of learning from any kind of experience that is not authorized by the dictates of their creed.

Extreme examples of dogmas are found in the most virulent forms of nationalism, such as German National Socialism and the imperialism of Japan during World War II. It is seen in religious intolerance and rivalry from the time of the Crusades down to the present day. Even the fanaticism of some sports fans, who consider a loss by their team to be a catastrophic injustice, has the flavor of dogmatic belief. Of course, dogmas can develop anywhere some group of people decides they already know the truth, whether they have reasonable evidence or not. Allegiance to a dogma invariably implies a total commitment to some kind of authority, be it religious, political, or scientific.

This is precisely where scientific materialism and science part company, or at least where they should. Science has aspired to discover truth through the dispassionate, objective examination of data and theories. Dogmatic beliefs, which historically hindered our search for truth, were to be replaced by the working hypothesis—a tentative belief based on empirical evidence and reasoning that would guide research. A critical attitude was essential to this style of inquiry. Recall the words of Richard Feynman: "Experimenters search most diligently, and with the greatest effort, in exactly those places where it seems most likely that we can prove our theories wrong. In other words we are trying to prove ourselves wrong as quickly as possible, because only in that way can we find progress."[3]

This is just what happened at the end of the nineteenth century when the principles of scientific materialism appeared to reign triumphant. Despite widespread confidence among many scientists that all of the major riddles of the universe had been solved, science did not rest on its laurels. It remained critical, curious, and open-minded enough to prove its flawed theories wrong. Planck, Bohr, and Einstein made key discoveries that showed classical physics to be limited when it came to the atom, subatomic particles, gravity, space, and time. Unfortunately, quantum physics, relativity, the Standard Model describing the subatomic world, along with chaos theory, superstrings, and the like are not easily absorbed into common sense, especially when common sense has for so long been molded by scientific materialism. As we have seen, Einstein himself, believing that the whole of science was "nothing more than a refinement of everyday thinking,"[4] could not accept all of the implications of the new physics. The quantum world is anything but a version of everyday thinking as it has so far existed.

WHERE'S THE SCIENCE?

Suppose that archeologists in some far distant future, sifting through the relics of our time, happened upon the remains of a nuclear power plant. If these future investigators knew about twentieth-century physics, they would understand that the scientists of our time had used relativity and quantum physics to devise the plant's nuclear reactor. In addition, if they wished, they could conceivably replicate a twentieth century nuclear power plant from this scientific information. However, if these archeologists had only gathered information on nuclear energy as explained in the popular media of the twentieth century, they would be faced with an interesting paradox: The popular story of the atom is drawn from outmoded classical concepts, providing a description of nuclear energy that is essentially a fairy tale. No working nuclear reactor could be built from such information.

What, for example, does popular journalism—be it television, radio, or the print media—tell us about the workings of nuclear reactors or the atomic bomb? The key word in the standard explanation of nuclear fission is "chain reaction." It is explained that highly radioactive materials such as uranium and plutonium emit substantial numbers of subatomic particles. Some of these particles are capable of splitting the nucleus of uranium or plutonium atoms, thereby releasing a tremendous amount of energy. In nuclear reactors and atomic bombs, special conditions are set up so that the number of free subatomic particles increases to the extent that a chain reaction occurs, producing heat in the reactor and an explosion in the bomb. For public consumption that's as far as it goes.

This explanation amounts to a purely mechanical image: small billiard balls are striking larger billiard balls, making them break apart. Nowhere is it mentioned that, in commonsense terms, the physics leading to atomic fission can barely be imagined (in the case of relativity) and is impossible to imagine (for quantum physics). The principles that led to this important discovery, so influential on modern life, are not even alluded to. Therefore the possibility of our developing a new kind of common sense, or a new way of imagining—one that could give us a more accurate or up-to-date picture of the universe we are a part of—is not explored. The public perception of the atomic world is dictated by a mechanical understanding that is now completely outdated.

Journalists and authors may complain that the mathematics and other technical issues basic to modern physics make it impossible for them to go deeper into these matters. Perhaps. However, the workings and implications of today's physics can be discussed *if* the writer is willing to confront the classical explanations in which we are presently mired. Take our quantum example of the superposition, where a rifle is aimed in a random fashion at a concrete wall. One implication of a "cloud of bullets" emitted from the rifle with its "wave form collapsing" to a single point on the wall is that the universe is not made of hard stuff that interacts mechanically. Moreover, in our examples of quantum entanglement or nonlocality, there is the strong suggestion that the mind itself is intimately involved in what were once considered purely physical processes. Indeed, there is very strong scientific evidence for these "impossible" situations that contradict common sense. The basic theories of quantum physics do not speculate beyond the data. However, if one is biased, even unconsciously, by the closure principle (that all causes are physical) and the other beliefs of scientific materialism, one will not seriously attempt an explanation of quantum relationships— essentially because one doesn't believe such things are real. This is a result of dogmatic thinking.

To their credit a small number of journalists have been willing to explore some of these implications. Bill Moyers, in his PBS television series "Healing and the Mind," presented examples where mental, non-material factors have had proven, positive effects on physical health.[5] These include practices such as meditation, acupuncture, and herbal medicine. He even showed a segment where a Chinese marshal arts master was able to fend off an attack by one of his students without any physical contact, apparently by some kind of invisible energy. Whereas scientific materialism rejects such reports, modern physics (and, as we shall see, Eastern philosophy and practice) allows for their possibility. The only problem here is that the media have been largely unwilling to explore this.

FOREGONE CONCLUSIONS

Nowhere is this limitation more evident than in popular explanations of research on the mind. Here the danger is double because, as we have seen, current research in cognitive science is often guided by the outdated principles of scientific materialism. So not only are journalists likely to use examples and metaphors based upon unverified conclusions, but the researchers themselves are equally tempted. One case in point is a *Scientific American* article entitled "How the Brain Creates the Mind," by Antonio R. Damasio, a leading neuroscientist and frequent contributor to news articles on the mind and brain.[6] The first sentence of this article reads: "At the start of the new millennium, it is apparent that one question towers above all others in life sciences: How does the set of processes we call mind emerge from the activity of the organ we call brain?" It is important to notice from the outset that the fundamental question "*Does* the brain create the mind?" is not seriously advanced. The author does address philosophical objections to the purely physical view of the mind, but in such a cursory manner as to give the impression that there are no promising, viable alternatives. So if Dr. Damasio takes it as a foregone conclusion that the physical brain creates the mind, there must be strong indications from the scientific evidence that this is so.

However, as one reads his article, it quickly becomes apparent that no such scientific evidence exists: "The current description of neurobiological phenomena is quite incomplete, any way you slice it," he says. One of the major problems here, according to the author, is "the large disparity between two bodies of knowledge—the good understanding of mind we have achieved through centuries of introspection and the efforts of cognitive science versus the incomplete neural specification we have achieved through the efforts of neuroscience." Yet we are not to worry, because Damasio is sure that neurobiology can bridge the gap sometime in the future. Those who would question the lack of current evidence are cast as naysayers, jumping to premature conclusions.

Later in the article the author attacks the central problem: *How* do physical brain structures either create or equate to consciousness? That is, how does a bunch of electro-chemical circuitry create our personal experiences of consciousness? Damasio says that consciousness developed naturally as part of humankind's biological adaptation through evolution. Human consciousness was required for survival of our species. It evolved so that we could keep track of our performance, test our hypotheses, survive challenges, and advance. Where did this consciousness arise? He refers to areas in the brain's core that he believes function to map our activities. These neural structures create "second order representations" of what we are doing at a given time: two systems are operating, one of which is watching or "mapping," the first. While we are engaged—say, in reading or running—there is a built-in sensation of our being conscious of our activities, a kind of echo, which generates the sense of an "I." If this is the case, the duality between the physical brain and the nonmaterial "mind" that thinks, imagines, perceives, etc., is done away with. The mind comes *as* the brain and not separate from it.

However, this theory turns out to be no more than a clever restatement of the problem. Damasio begins with a physical premise, defining mind as "a set of processes." But if consciousness is central to that which we call mind, what scientific evidence demands that we characterize consciousness as a set of processes? Aren't we just as likely to say that we are normally *conscious of* various processes—such as our thoughts, emotions, and perceptions—that comprise our "mind"? Moreover, does this dualistic definition of consciousness, this echo-like sensation, include the whole variety of experiences we call mind or consciousness? Often we are indeed "aware of" some phenomenon, but aren't there also times when we are simply "aware"? Although such moments may be rare, they do occur, as for example when we are unexpectedly confronted with a completely novel situation. Suddenly we trip and fall, or we are awakened from sleep by a loud noise. For a split second, or perhaps longer, we are intensely conscious but with no object of our awareness—even of ourselves. This is a nondual form of consciousness. Later we might say we were "disoriented." As we shall see in part two, nondual consciousness has been extensively explored by several contemplative spiritual traditions.

The picture Damasio paints is a purely physical one: electrochemical reactions between two systems, one of which—the one supposedly presenting "second-order representations"—acts as a kind of overseer or meta-program. Yet just how do two, or even a dozen, simultaneously operating neural systems create the consciousness or feeling of our subjective experiences? Similar programs, though of lesser complexity, exist in computers. Is your computer's operating system "conscious" of the applications it runs? Do Linux, Windows, or MacOS have a first-person experience of the activities of Netscape or Internet Explorer? Could any computer, no matter how complex, ever be shown to be conscious simply because it has an operating system designed to oversee its activities? Or does Damasio's view imply that we are as unconscious as computers?

The heart of the problem goes back to the word "representation." The very notion of representations (in terms of images, symbols, examples, characterizations, agents, etc.) occurring *in* neural structures is only hypothesis, a conjecture meant to bridge the gap between mental consciousness and brain matter. It does happen that brain scans sometimes show visual patterns of brain activity that roughly correspond to certain primitive visual patterns as they are seen firsthand. But these limited correspondences can be misleading. When we see blue, for instance, no part of the visual cortex turns blue. When we smell the fragrance of a rose, no neurons take on that fragrance, nor are there literally any sounds in the brain when we hear a friend's voice. None of the sensory experiences, or *qualia*, have any physical attributes, such as mass, spatial dimension, or velocity, so there is no good reason to imagine that they occupy any region of physical space, such as the brain.

All that neuroscientists really know is that specific neural events contribute in some manner to subjective experiences or representations. The fact that a brain injury or electric stimulation of some area of the brain can cause a speech defect or some other experience in a human subject is no proof that the neural effects are either equivalent to or are solely responsible for the creation of actual, personal, first-person experience, let alone consciousness itself. Nor does it explain how that process operates.

If some kind of "mapping" is occurring, exactly how is that map drawn as a conscious entity? We are not talking about roadmaps here. We are talking about maps that are themselves our perception of the quality we call consciousness. No one has ever empirically demonstrated or presented any cogent argument for concluding that such subjective experiences are actually located in the brain or are equivalent to any neural structures. Nonetheless, neuroscientists generally believe that any mental event that is taken to be "real" must be lodged in the brain. They draw pictures that conform to common sense—often quite convincing and ingenious pictures—but no more valuable for understanding consciousness than depictions of the brain as a hydraulic system or telephone switchboard. Once again this is simply an expression of the metaphysical assumption that if something exists, it must be physical.

Dr. Damasio's argument has other significant weak points, faults he shares with the writings of many of his colleagues in neuroscience. His reference to "the good understanding of the mind we have achieved through centuries of introspection," is puzzling. There have only been two significant attempts at a science of introspection in the West: among religious contemplatives—whose experiences have had only a slight influence on Judaism and Christianity and virtually none on science—and the fledgling and discredited introspectionist school of psychology that existed for a few scant decades at the turn of the twentieth century. If Damasio is referring to the observations of western psychology, philosophy, literature, or the social sciences, nowhere is there *less* consensus when it comes to understanding the mind. Freudian psychologists disagree with Jungian psychologists, phenomenological philosophers disagree with existential philosophers, and novelists and social scientists agree on very little at all. There is no "good understanding" of the mind in the Western scientific tradition based on introspection.

Therefore, on both sides of the argument—understanding of subjective experience, as well as "the incomplete neural specification we have achieved through the efforts of neuroscience"[7]–there is no basis for confidence that science has come to a satisfactory understanding of the mind as a product of the brain or that such an approach is promising. The giveaway in articles such as this one is the degree to which hope is placed in future research rather than upon concrete evidence. After expressing admirable skepticism that he would be "foolish to make predictions" about the content of future discoveries, Dr. Damasio boldly states his belief that "it is probably safe to say that by 2050 sufficient knowledge of biological phenomena will have wiped out the traditional dualistic separations of body/brain, body/mind, and brain/mind."

The manner in which such beliefs guide ongoing research is demonstrated in a more recent *Scientific American* article on neural encoding in rats:

> Exactly how the brain translates a language of electrical pulses
> into such fine and varied perceptions has long been a pro-

found puzzle and one of the holy grails of brain research. *To crack the neural code is to open the doors to comprehending the essence of who we are* [our emphasis]. Our abilities to speak, love, hate and perceive the world around us, as well as our memories, our dreams, even our species history, emerge from the combination of a multitude of tiny electrical signals that spread across our brains, just like a thunderstorm sweeps the sky on a summer night.[8]

SCIENCE OR PROPAGANDA?

With such untested assumptions guiding cutting-edge neuroscientific research, their appearance in the popular media is only to be expected. Antonio Damasio, for example, is a member of the prestigious American Academy of Sciences and directs the Brain and Creativity Institute at the University of Southern California. That makes him a prime source for reporters writing on this topic. But he is by no means alone as a major influence on public perception of the mind and brain. A number of cognitive scientists and philosophers of mind hold similar views and are authors of, and are frequently quoted in, articles on the mind and brain in the popular media. In doing so they present an image of consensus about the mind/brain problem that has gone a long way toward convincing the public that these purely speculative views are scientific facts. They are today's version of the X Club. Here are some examples of this influence with their respective errors unmasked by a more skeptical viewpoint that asks for scientific evidence.

From "How to Think About the Mind" by Steven Pinker in *Newsweek*: "Modern neuroscience has shown that there *is* no user. 'The soul' is, in fact, the information-processing activity of the brain. New imaging techniques have tied every thought and emotion to neural activity. And any change to the brain—from strokes, drugs, electricity or surgery—will literally change your mind."[9]

However, the existence of new imaging techniques in no way makes it a "fact" that the soul (or anything else considered part of the mind) is "information-processing activity of the brain." These new images merely show activity in the brain that can be correlated to behavior and mental experiences. The images do not show that mental activity emerges from or is created solely by the physical brain.

From the Public Broadcasting System's five-part series, *The Secret Life of the Brain*:

". . . but as neurologist Carla Shatz says, 'There's a great mystery left. Our memories and our hopes and our aspirations and who we love, all of that is in there encoded in the circuits. But we only have the barest beginnings of an understanding about how the brain really works.'"[10]

Well, if we have only the barest beginnings of an understanding about how the brain really works, isn't Dr. Shatz stretching a bit to conclude that "all of that" (memories, hopes, aspirations, love) "is in there encoded in the circuits"? And just how is love or hope encoded?

"Emotions are not the intangible and vaporous qualities that many presume them to be. Brain systems work together to give us emotions just as they do with sight and smell."[11]

Here Dr. Damasio (speaking on Dr. Shatz's broadcast) is assuming that emotions are precisely defined and well understood by neuroscience—which they are not. How, for example, do emotions and consciousness (another poorly understood term) overlap? Such a lack of precise knowledge renders comparisons with the operations of the senses mere speculation.

From "Just Say Om" by Joel Stein in *Time*:

"By shutting down the parietal lobe, you can lose your sense of boundaries and feel more 'at one' with the universe, which probably feels a lot less boring than it sounds when you try to tell your friends about it."[12]

Here, the fact that meditators' parietal lobes shut down doesn't mean that this is the cause for their losing their sense of boundaries and feeling more at one with the universe. This may only be a complementary effect, like the sweat produced during physical exercise. Indeed, these sensations may not be the principle aim of these practices. Much more may be happening. It is unlikely that any meditation master instructs his students: "Now, shut down your parietal lobes."

Unsubstantiated conclusions of this sort are also transmitted to students in public school and college textbooks. The belief that mental phenomena must be products of matter—the brain—is built upon the following widely propagated interpretation of evolutionary theory: (1) after the geological formation of the Earth, there existed on our planet only nonliving molecules; (2) the earliest forms of life, bacteria, emerged from these molecules; and (3) evolution proceeded from bacteria, through successively more complex life forms, until humans finally evolved. According to this argument, if life emerged from inorganic matter, then so has consciousness. Although science has ample means of detecting inorganic matter such as molecules, and describing their formations and interactions, it has absolutely no instruments for detecting consciousness. If we were to use the instruments and methodologies of physical science as our only basis, we would not know that consciousness existed at all, anywhere, any time. So here a set of analogies is substituted for the rigors of science. Yet virtually all major college textbooks, supposedly imparting scientific knowledge to students, follow this speculative line of reasoning.

Take for example John Anderson's *Cognitive Psychology and Its Implications*, which follows the familiar information processing approach to the mind and brain.[13] He describes visual perception in the following manner: Receptors in the retina of the eye transform light energy into neural information. Here, light is converted into neural energy by a photochemical process whereby non-conscious cells "detect simple patterns of spots of light and darkness in the visual field." Now, although light may indeed be converted into neural, electrochemical energy, this does not account for the quality of light and dark that is perceived, a very important aspect of visual perception. Such low-level cells may indeed react to pulses of energy, just as a photocell may react to a passing shadow and trigger a door to open. But there is no evidence that in either case experiential patterns of light and darkness are detected. (Detection implies a quality of awareness. In contrast, the legs of a frog, separated from the rest of its body, will still *react* to an electrical stimulus.) The qualities of the visual world that we humans consciously detect are far more subtle and varied than a mere unconscious reaction to patterns of black and white. Do such low-level cells really detect a magnificent sunset or a dizzying height? In fact neuroscientists generally agree that there is no one-to-one correspondence between the frequencies and intensities of light and the multifaceted qualities of our visual world.

Anderson jumps to similar unproven conclusions in his description of human cognition, which he says is achieved "by patterns of neural activation in large sets of neurons." While saying that the brain "encodes cognition in neural patterns," he admits that it isn't known *how* this happens.[14] Such statements may sound like facts, but they are simply speculation. And, opining that computers may be able to duplicate the mental phenomena of humans, the author further declares that "computer systems . . . have been shown to be capable of . . . displaying frustration." This feeling of frustration occurs in "large patterns of bit changes." One could easily imagine that a rapid frazzle of bit changes is somehow similar to human "frustration." But this is no more convincing than declaring that crosswinds roiling up the waters of a lake mean that the lake is frustrated. These are just analogies suggested by common sense—a misguided common sense based upon the beliefs of scientific materialism. Nowhere does Anderson present evidence of this emotion in the computer nor does he show how feeling can become embedded in patterns of bit changes.

TO SEE THE WIZARD

At the end of the movie *The Wizard of Oz*, Dorothy and her three companions have finally reached the emerald city and have been granted an audience with the powerful Oz. They are ushered into a cavernous hall where they experience the frightening presence of a monstrous image. Flames belch forth and a scowling face intimidates them. However, Dorothy's little dog, Toto, pulls aside a curtain revealing a harmless little man working a control panel. "Pay no attention to the man behind the curtain!" growls the image, but the game is up. The false image disintegrates because Toto was skeptical. And given a dose of skepticism—

the kind science is supposed to rely upon to fulfill its mission—those arguments coming from neuroscience that rely on the beliefs of scientific materialism look pretty weak. To review:

(1) Consciousness, the basis for all of our mental experiences—of feelings of "I," the experience of thought, emotion, sensation, and so on—is neither well defined nor well understood in the cognitive sciences. It cannot even be detected with scientific instruments.

(2) Lacking a precise understanding of the first-person experiences of consciousness, attempts to equate them with the brain's physical structures are premature. Although correlates between brain functioning and *behavior* have been discovered, no one has explained what it is about neural systems that could enable them to generate or even influence conscious, first-person *experience*. Unfortunately, mere speculation has been offered as scientific fact.

(3) Descriptions in the cognitive sciences of how the mind "emerges" from the physical brain are incompatible with any other kind of emergent property observed and described in physical sciences, such as chemistry and biology. In fact, the emergence of the mind from the brain is not observed at all.

(4) Alternative explanations, such as William James's suggestions that consciousness and mental experience may be filtered by, rather than produced by, the brain (in the manner that a prism affects light), are not considered, even though this thesis is completely compatible with scientific research. Only physical explanations have been investigated, based on the assumption that mental events must be either physical or nonexistent.

(5) "Future research" will supposedly confirm today's unproven conclusions. The boldness of materialist assertions in the cognitive sciences shows a disregard for the scientific method and points to the immaturity of this young science.

There is a fundamental flaw within the physical approach itself: Scientific materialists consider organic, biological events (such as photosynthesis or cell mitosis) to be emergent properties of inorganic chemical processes (chemical elements, compounds, etc.). These, in turn, are emergent properties or functions of configurations of atoms and elementary particles. According to physics they are the ultimate physical basis of the universe and thus necessarily the basis of the whole scientific materialist belief system. However, as we have seen, physicists describe matter and energy in terms that are so abstract, insubstantial, and even non-local that they appear more like conceptual constructs, more like possibilities than anything solid and "real." So if the matter itself in *material*-ism is so vaporous, does scientific *materialism* have solid ground beneath its feet?

TRUTH OR CONSEQUENCES

At the end of his article quoted above, Antonio Damasio writes: "By understanding the mind at a deeper level, we will see it as nature's most complex set of biological phenomena rather than as a mystery with an unknown nature."[15] This is an extremely significant pronouncement, echoing the early debates between science and Christianity. Although Dr. Damasio may not intend it as such, it is essentially a religious statement revealing the fundamental belief and the essential object of worship of scientific materialism.

"Religion" can be defined as "belief in a divine or superhuman power or powers to be obeyed and worshiped as the creator(s) or ruler(s) of the universe."[16] Scientific materialism has gotten rid of God and mystery, but still hangs onto the belief in a superhuman power—*nature*. If the mystery of the mind is lost, along with all of the spiritual beliefs found in the mind, we can still marvel that it is "nature's most complex set of biological phenomena." Damasio does not leave us with atheism (a denial that some God exists), nor with agnosticism (that we cannot know whether or not some God exists). Nor are we left with the openness and intellectual freedom of a working hypothesis, of a faith in exploration. Rather, we are given a metaphor: the goddess, Nature, described as a marvelously complex though ultimately intelligible source or *creator*. Yet what precisely is this goddess, Nature, and where can she be found? Floating behind the veil of human perceptions, metaphysically real, she is the essence of physical existence.

Thus scientific materialism is essentially religious both in its underlying tenets, abstracted from Christianity, and in its belief in physical nature as the sole creator and mover of the universe. Nature—as the Big Bang, the chemistry of the genome, subatomic physics, and everything yet to be discovered by science all rolled into one—replaces God. The fact that nature isn't depicted in anthropomorphic form as Mother Nature (instead of the bearded gentleman of naïve Christianity) doesn't weaken her position. In this belief system, nature is the sole "unmoved mover." Though scientific materialism may not resemble the more familiar traditional religions, this does not change the fact that this dogma, wearing the mantle of science, exerts a powerful and largely undeserved influence on society and its beliefs. And in fulfilling T. H. Huxley's dream of replacing religion, scientific materialism has become, in effect, the official religion of modern nations.

AFTERSHOCKS—SCIENTIFIC MATERIALISM AND SOCIETY

Imagine a teacher, say a Hindu, in a U.S. public school classroom telling his students that the story of creation in the Bible was only a myth and that the miracles attributed to God and Jesus never happened. Then suppose that in another public schoolroom a Christian teacher said that reincarnation and karma were false doctrines. And in another public school a teacher who is a Muslim states that the Jews are not God's chosen people. In the United States with its separation of church and state, it would be illegal for any teacher to make these negative assertions in a public classroom even if the teacher wasn't promoting his or her own religion. However, advocates of scientific materialism make all of the above claims in public classrooms, explicitly or implicitly, at the same time that they promote their own belief systems. Dominated by the dogma of scientific materialism, classroom science denies without argument the miracles, worldviews, and cosmologies of religion. Yet the fact that many currently accepted scientific theories include nonphysical possibilities is often overlooked. This omission betrays the open spirit of investigation that is the hallmark of true scientific inquiry. By limiting science in this way, researchers are ignoring fields of inquiry that could lead to discoveries that lie beyond our present knowledge or imagination.

In capitalist-oriented countries such as the United States, science receives a large slice of national budgets. Government-sponsored basic research drives the development of new technologies, which in turn create great wealth. Weapons development and medical research are well funded. This technological orientation requires a massive political mobilization, which is influenced and often guided by university science departments and research centers or by government and private institutes. These are the voices that are listened to in Congress and the White House when policy decisions regarding science are deliberated, and for the most part, these institutions toe the philosophical line of scientific materialism.

This problem is not limited to the United States. Scientific materialism permeated the Marxism of the former Soviet Union and China, where it has undeniable links to the repression of religions (alternate and competing belief systems), such as Christianity, Judaism, and Buddhism. The result, however, is similar whether the pressure of conformity is harsh and stems from an oppressive government or it is a more subtle combination of economic, political, and educational influences: scientific materialism dominates.

What are the consequences of this closed-minded attitude toward spirituality and the distortion of the true spirit of scientific inquiry and many of its discoveries? In the United States, surveys demonstrate that a majority of the population claims religious affiliation; most Americans believe in a nonphysical deity. Is it any surprise, then, that many of them, when exposed to the very limited view of scientific materialism in the guise of "science" are put off? Scientific materialism tells them that their very minds are nonexistent or are nothing more than meat. It denies that their personalities, intuitions, and spiritual experiences are anything more than ghostly secondary effects of the interactions of biological hardware and software.

In addition, what of those students who do take an interest in science, believing that the practice of science follows the open-minded, exploratory spirit of the scientific method? They study textbooks that either imply or boldly declare that as-yet-unproven theories are definitely true or will certainly be proven true in the future. They are exposed to an attitude toward science that promotes conformity to the foregone conclusions of scientific materialism even as it pretends to favor free inquiry. Those people who see the contradiction are left with the choice of buckling under or striking out on their own. Alternatively, they may become discouraged with science altogether and choose another career.

The consequences of this narrow view extend to other vital areas in society. The U.S. Constitution forbids the promotion of religion in public schools ("Congress shall make no law respecting an establishment of religion."). This has come to be interpreted in such a way that any kind of education about religion must be subjected to the kind of "objectivity" dictated by scientific materialism. Religious studies courses in colleges and universities frequently require that students read scientific-materialist authors such as Emile Durkheim, Marx, and Sartre, along with postmodern thinkers whose views are often versions of materialism. Adoption of the principles of materialism and reductionism promoted by these detractors of spirituality is commonly viewed as an objective outlook, a proper educational approach to the study of religion, whereas traditional spiritual views are seen as subjective and therefore suspect. The dice are loaded. No one questions what this "objective outlook" really implies as we have done throughout this first part of this book.

Due to this biased attitude, Americans remain largely ignorant of all but their own personal spiritual beliefs—and this in an age when religion and politics have become thoroughly mixed on the international stage. There seems to be a misplaced fear that if, say, a religious believer were brought into a classroom to explain why he believed in his creed and what his religion meant to him personally, this would appear to be promoting religion. The U.S. Constitution originally forbade the promotion of religion in order to avoid the creation of an official state religion such as the Church of England. Its intent was not to stifle spirituality. Forcing all the children in a classroom to recite or listen to the Lord's Prayer each day would indeed be the promotion of an official religion, but allowing children to describe their own spiritual beliefs, or lack of belief, and to be exposed to other beliefs does not promote any particular religion. It simply opens up the mind to wider possibilities and a broader understanding of the world at large. Isn't it ironic that scientific materialism acts like a guardian of objectivity (defined on its own terms) when it is itself a religious dogma? Wouldn't the public be better served by classroom presentations of spiritual beliefs from both the "objective" and subjective-personal points of view? Is that not the kind of objectivity we need? It is certainly closer to the true spirit of scientific inquiry.

THE CENTURY OF SCIENTIFIC MATERIALISM

As we have seen, many of the beliefs of scientific materialism have been deeply challenged, if not overthrown, by advances in modern physics. It seems that this was inevitable. The basic tenets of scientific materialism were inherited from Christianity in its desire to understand the mind of God through its reflection in the physical universe, "the heavens." The assumption of a purely physical universe—elaborated through objectivism, metaphysical realism, physical reductionism, the closure principle, and universalism—was then logically extended to the conclusion that God, gods, demons, miracles, and any other nonmaterial reality was illusory. The Christian dogma that the universe was governed by a single "personality"—God—evolved into a dogma that focused on an impersonal, physical nature based on chance and necessity that runs the show.

In this story of ironies and contradictions, the very tools that ousted God from preeminence—the empirical scientific method based upon theory and experiment—finally exposed the failure of scientific materialism and the limitations of classical science. The process described by Richard Feynman—science striving diligently to prove its faulty theories wrong—eventually caught up to the mistaken assumptions of scientific materialism. The reintroduction of philosophical debate on the issues of scientific realism, instrumentalism, underdetermination, cultural paradigms, and historical shifts in scientific viewpoints further opened the scientific mind, at least in physics, to broader possibilities.

Even so, the influence of scientific materialism on both scientists and the public was greatest during the twentieth century, and its dominance increases in the twenty-first century. The modern world came to be called "the scientific age" as numerous discoveries and inventions transformed society from top to bottom. There were great benefits, especially in health care, transportation, agriculture, industrial production, and communications. But the twentieth century was also the period of human beings' greatest inhumanity to one another as well as unprecedented degradation of the natural environment. During that century, the most murderous despots in history, including Stalin, Hitler, and Mao Zedong, were able to enlist the services of scientists in supporting their crimes against humanity. Indeed, numerous scientists and engineers throughout the world have applied their knowledge and ingenuity to develop weapons of mass destruction that now terrorize humanity as a whole.

Although many scientists abide by high ethical standards, scientific materialism gives them no incentive to do so. A mechanical, impersonal clockwork universe makes no reference to ethics or virtue. Indeed, reducing human subjectivity, the human mind, to neural processes operating according to the impersonal laws of physics (which imply neurobiology, chemistry, and so on) undermines any sense of moral responsibility. Given such an amoral background, it is not surprising that crime, selfish attitudes, and public indifference to the plight of the less fortunate have come to dominate modern life.

Even in health care, the purely physical approach has drawbacks, since it often attacks the symptoms of disease but not the cause. Moreover, we do not know the specific causes of many illnesses, both physical and psychological, especially chronic diseases such as hypertension, chronic back pain, certain forms of cancer, and many others. Some of these causes may be of a psychological nature, requiring treatment of the mind through nonchemical means. Scientists are finding direct links between stress, inflammation, and disease. Furthermore, traditional approaches do not always work. For example, drug therapy for attention deficit disorder effectively suppresses the symptoms of this problem in only about 50 percent of cases, and the widespread use of antidepressants among children has been linked to suicide. The discovery of dangerous side effects of pharmaceuticals is a common occurrence.

The materialist approach to medicine has led to the desire for a "quick fix"—just pop a pill and let chemicals take care of it. Drug, tobacco, and alcohol addiction follow the same logic. There may be more to mental and physical illness than just chemicals, but the physical bias of scientific materialism has largely marginalized alternative therapies that show promise. The slow, grudging acceptance of acupuncture is just one example of this. For example, even though the "channels" used by acupuncturists, shown vividly on their charts, have not been detected by modern anatomy, acupuncture is often effective in cases where physical medicine has failed. There is also a growing public interest in herbal remedies, whose curative influences may take longer than manufactured pharmaceuticals but which also may not have as many troublesome side effects as the latest drugs. Finally, although it has long been known by clinicians that the positive attitudes they may express toward their patients (such as genuine warmth, concern, and confidence) affect healing, this has only recently been incorporated as part of the curriculum in medical schools, and still only to a very limited extent.

T. H. Huxley and his modern counterparts expected that their brand of scientific philosophy would eventually wipe out traditional religion. But for a number of reasons this has not occurred. To be born for no reason and then be extinguished forever is a disheartening and perplexing picture of human life. For some people a retreat from this meaningless view of existence is undoubtedly their main spur toward religion. Others may recognize the wisdom stored in spiritual traditions or have an intuitive sense that there is more to life than materialism's bleak prospects. As is abundantly clear in the examples given above, the reasoning of scientific materialism is often faulty and outdated. Accepting its conclusions on authority may not be wise. Perhaps we still have access to a greater common sense than the assumptions and mental habits we have inherited from this dubious dogma.

Such a viewpoint does not require a negation of science. The open-minded, exploratory, unbiased ideals of science have already identified many of the nonphysical aspects of the world we inhabit. To live honestly and authentically with the hypotheses of science does not mean we must give up all of our beliefs in the miraculous or the mysterious. Rather, there is growing evidence that science and religion may complement each other—not in the sense of keeping rigidly to their traditional spheres, but in collaboration toward a wider and clearer understanding of life and the universe.

To test this hypothesis, we need to reintroduce a spirit of true empiricism into the study of both science and religion. Students should be introduced to the ethical, historical, and philosophical contexts in which the sciences and religions of the world have emerged and coevolved throughout history. Furthermore, they should be encouraged to extend the empirical quest into the inner realm and to get to know their minds firsthand, utilizing that knowledge to fathom the import of both science and spirituality.

"Know thyself," the ancient call to self-observation, holds the promise of genuine happiness for us as individuals. Indeed, it may also lead to knowledge and wisdom embracing life itself. Without intimate knowledge of our subjective nature, how can we truly follow the dictum "to thine own self be true"? Self-knowledge clarifies experience, leading to maturity. In addition, as we shall see in part 2, the most subtle practices of introspection may reveal a surprising complementarity between the mind and the universe.

Science, for its part, has the potential to provide a wonderful bridge for interreligious dialogue and for comparative religious studies while helping to sharpen philosophical reflection and deepening our appreciation of these received wisdom traditions. In this regard William James wrote, "Let empiricism once become associated with religion, as hitherto, through some strange misunderstanding, it has been associated with irreligion, and I believe that a new era of religion as well as philosophy will be ready to begin."[17] Let the twenty-first century be one of cooperation between spirituality and science and perhaps some of the causes of the strife and misery that pervaded the twentieth century can be avoided.

PART TWO
Consciousness—Completing the Picture

WE'VE SEEN HOW and why the mind, along with its essential quality, consciousness, was left out of the scientific picture of the universe. This has caused problems for science, for spirituality, and for our social well-being. However, physics, the most mature of all the natural sciences, has been forced by experimental evidence to include the role of the observer as a complementary aspect of physical reality. Mind and matter seem to form an inseparable pair, in essence an interconnected universe. Meanwhile the other sciences, supposedly basing themselves on the conclusions of physics, still equate the mind with the brain—a purely physical reality, if even that. Cognitive scientists in particular remain entranced by the illusions of scientific materialism, some no doubt dreaming of the day when a sophisticated computer will actually "become human," or at least conscious.

Certainly we cannot place all of the blame for humankind's greed for material plenty and power, nor its blind destructiveness, on scientific materialism. Yet with few exceptions, the common-sense logic that guides many of our thoughts and actions and much of public policy is strongly influenced by the tenets of that philosophy. This reinforces in us a belief in the ultimate value of the material realm, whether in reference to our physical health or the choice of military solutions to diplomatic problems. If our physical existence is all there is to it, we have little incentive to probe elsewhere or act otherwise. There are many altruistic people who by nature, education, or from personal experience are open to the plight of others and who lend a hand. Spiritual traditions of course encourage such actions, basing their altruism on truths thought to transcend the physical plane. Those who follow such moral codes without the support of spiritual beliefs do so on their own, going against the grain of logic based on popular materialistic thinking.

We have suggested that accepting the mind as more than merely physical and exploring it through introspection—not with scientific instruments alone—could balance our lopsided view of reality. So far we have mainly argued in the negative, that leaving the mind out of nature has been a catastrophic mistake. Before demonstrating the positive effects of accepting, exploring, and developing the mind, which is the forte of Eastern religions and philosophies, we will argue that science and Western philosophies have themselves brought us to the brink of a new paradigm or worldview. Their discoveries have placed us in a position where the contemplative insights of the East may become both acceptable and welcome.

Coming Full Circle
And Meeting the East 6

ONE THING should be obvious: without our minds we wouldn't know anything. The very act of knowing implies some process of recognition, some awareness that "knows." In the broadest sense we could define mind itself simply as a *knower*. The experience of touch provides an example of the kind of very basic awareness being referred to. When our finger makes contact with some object, say the water in a glass, there is an instant when we know we have touched the water. This bare recognition of contact—perhaps just the awareness of change—is quick and simple, coming before we note such things as the wetness, texture, temperature, or even identify that what we have touched is "water." That consciousness is an essential property of the mind. Conscious awareness brings us into contact with everything we know, from the world of the five outer-directed senses to the sixth sense of mental perception and its associations with the inner world of thought, imagination, emotions, and so forth.

From that viewpoint there would seem to be nothing we know more intimately than our minds. We are constantly engaged in this process of knowing—swimming in it. Therefore, if we can be sure of anything existing, we can be sure of the mind. Yet what exactly is it? Where is it? What are its proportions? Strangely, when we begin to explore all of the intricacies of our mental lives, the mind seems to be a part of everything, like a mirror that has broken into a thousand pieces reflecting many things from many angles. At that point our mind appears to slip through our fingers. We cannot grasp just what mind is. Like a ghost, it loses definition, fading in and out of view. Then, since it reflects all the richness and complexity of our existence, we tend to identify mind with its reflections rather than the awareness that makes them visible. Perhaps the mind is so slippery simply because it is so close, so obvious, hidden in plain view. It's like looking for our glasses or our hat when we've been wearing them all along.

The difficulty in pinning down the mind is confirmed when we seek help from dictionaries. We find general definitions of "mind," such as "that which thinks, perceives, wills, etc; seat or subject of consciousness; the thinking or perceiving part of consciousness; intellect or intelligence; all of an individual's conscious experience." Consciousness is described as "the totality of one's thoughts, feelings, and impressions; conscious mind."[1] Such lists or collections of the contents and activities of the mind are useful in the general sense; however, we never experience such a collection that we can call "the mind" directly, in one bite. The mind is not like a collection of coins or shells that we can identify at a glance.

We've seen that in the modern Western world, with its Christian heritage and its science, the mind has been pointed outward to such an extent that it has virtually forgotten itself. Whether it sought the laws of the heavens reflecting the mind of God or simply the physical universe, the Western mind has wanted most to know what lies beyond. We have been so intrigued by physical objects that the mind itself, the seeker, became unfamiliar—a hidden phenomenon, a black box. When the mind did come into the picture, it was often viewed as a dangerous place harboring either sinful, demonic entities, or a murky pool of neurotic, subconscious impulses. Later, when science finally brought itself to study the mind, it was either equated with the familiar physical stuff science was accustomed to—the gray matter of the brain—or deemed nonexistent. So it is not surprising that the mind is still a mystery to us. Yet if the mind in the West tended to stare right past itself in its outer quest, in the East the general direction has been inward.

In fact, many of the spiritual and philosophical bases of thought in Asia begin with the mind itself, sometimes to the extent that—completely opposite to the view in the West—it is the mind that is the source of the physical universe rather than the other way around. Whether or not the many views of the mind in the East are proven or justifiable, this emphasis has certainly led to an intense exploration of the mind over several millennia.

COMING FULL CIRCLE 113

As we shall see, this exploration can in many instances be called a science. Although these traditions use different methodologies from those of Western science, their empirical standards are similar to those underlying scientific research. However, before we go farther in our examination, let's first explore in greater depth some of the implications arising from modern science as they relate to the nature of the mind and its role in human knowledge, for science itself has learned things that have a curiously Eastern ring.

A Fractured Fairy Tale

Science is nothing if not meticulous—it goes all-out. Stories of scientists working to exhaustion are commonplace. Whether motivated by intellectual curiosity, altruism, egotism, or greed, science has always been characterized by tremendous energy and determination. From all angles science is under pressure to "get it right" and push the envelope of human knowledge to the max. Despite the philosophical prejudices that have plagued science in the form of scientific materialism, it can nevertheless be said along with Richard Feynman that science tends to probe "most diligently, and with the greatest effort" into its own potential flaws and weaknesses in order to move forward quickly.

Given that, it is not surprising that instrumentalism arose to challenge the metaphysical claims of scientists. Instrumentalism is a kind of "show me" attitude. In a world where technological marvels derived from scientific theories have radically transformed modern life, it takes courage based on diligence to suggest that the theories themselves may nonetheless be inconclusive or false. Therefore, it is a sign of the thoroughness of science that it maintains such a high degree of skepticism as is found in the instrumentalist positions we viewed earlier.

Recall that instrumentalism exposes areas where science has stepped beyond the evidence to imagine metaphysical entities—supposedly real, independent causes operating behind the scenes. Such objective sources of phenomena are what realists crave to discover. Scientific realism assumes that when it discovers such laws, it can predict the outcome of experiments and devise new technology, and that these things happen because the laws are true. In opposing this view, instrumentalism doesn't make suggestions as to how things might function if those laws were not absolute truths; it merely advises science to be more cautious. The historic instability of scientific theories suggests that this skepticism may be wise.

Earlier we examined electricity and its centerpiece, the electron. Prior to Thomson's discovery of the electron in 1897, there were numerous theories about the nature of electricity. Eventually it was defined as subatomic particles (electrons) flowing from atom to atom. However, that theory was short-lived. We still use it to describe electricity to the general public and in beginning-level courses in physics and electronics. Here the electron has become a conventional truth—an instrument useful within specific contexts—just as was Benjamin Franklin's understanding of electricity when he experimented with lightning. The electron has instrumental value, because it has wide practical applications. However, after quantum physics and the standard atomic model came along, the electron was seen as part of a subatomic environment that is complex, fuzzy, and often controversial. The electron doesn't necessarily inform us about some ultimate reality.

Instrumentalism, or antirealism, has dogged scientific realism from Galileo's day up to the present. However, from the realist perspective that still dominates present-day "common sense," the implication that something can function perfectly well yet not be based on an absolutely "true" theory sounds unreasonable, even preposterous. Rather we assume that functioning things must reflect the truth. If not, how could they work? On what would their functioning be based? The only alternative explanation would seem to be magic. Yet, science in its search for causes based exclusively in physical reality discarded magic, demons, and miracles long ago. Any notion of nonmaterial causes became the great taboo— anything but *that*! Clearly there is wisdom in being skeptical of unproven claims. Otherwise, quacks and pseudoscientists could foist nonsense based on pseudoproofs onto the unsuspecting public. Yet, in banishing the nonmaterial did science go too far?

Underdetermination casts even more doubt on the foundations of scientific realism. If a theory is underdetermined, that is, if multiple theories can account for the same experimental data, that suggests multiple and contradictory underlying causes. The fact that this happens frequently makes one wonder if looking for underlying causes is even useful. These proposed causes, remember, are in the nature of mental constructs usually expressed mathematically. The hope here is that the mind that constructs these theories is capable of modeling reality, a reality that the minds of many scientists believe exists. However, do we know enough about the mind to justify that claim or, if that claim is reasonable, do we know enough to recognize ways in which the mind may deceive itself in its explorations? The instability of scientific theories could be a reflection of our limited knowledge of the depth and workings of the mind itself.

Quantum physics is yet another piece in this odd mosaic, this fractured fairy tale of scientific realism. Here we have mathematical formulas that work like a charm. But the reality they refer to is mind-boggling. Waves act like particles and particles like waves. Things lack certainty until measurements are made, but some vital measurements cannot be made at all. This inability to measure and determine the initial state of physical systems implies that systems as a whole are somehow indeterminate. Furthermore, events in one locality seem to be influenced instantly by events even light-years away, apparently defying Einstein's law that nothing can move faster than the speed of light.

The most revealing characteristic of quantum physics is the role of the observer in measurement: it is the act of observation, intimately wrapped up in the point of view of the scientist—his or her beliefs—that determines outcomes such as wave or particle and other physical states. It seems that at the subatomic level, the level that supposedly underlies all of physical reality, the mind acts as a potent, cooperative force in the creation of reality as we know it. Subatomic particles, the instruments that detect them, laws concerning their existence and expression, mathematics, and the mind all exist in dependence upon one another.

But What If They Are Right?

We have seen that modern neuroscience has for the most part followed the trend of scientific materialism in its attempts to find a physical basis for the mind. Behavioral effects resulting from alteration of brain functioning have been found and catalogued, but the key issue of the origins of consciousness has not been solved. Is the brain solely responsible for the generation of consciousness, or, according to one of William James's hypotheses, does it merely modify consciousness as a prism modifies sunlight? Is it both a necessary and sufficient cause for the generation of consciousness, or could there be other alternatives never even contemplated?

Let's suppose for a moment that the physical approach to the mind is correct—that the electrochemical brain indeed produces consciousness by itself, or *is* consciousness. That would make the phenomena that we call "our minds" pure illusions. Just as the scenes we view on a television screen or at the movies don't exist as such and are products of electronic and mechanical devices, our minds would be products of the brain. Given that, would complete information on the brain's functioning in relation to consciousness make the mind any less central as the primary instrument of human knowledge? Would it make our understanding of the mind any less mysterious from the inside—that is, how the individual must learn about, train, and use his mind? A competent mechanic can completely disassemble and then reconstruct a race car, gaining a perfect understanding of its functioning, but such knowledge would in no way guarantee his success as a racecar driver nor necessarily contribute in any way toward making him a safe driver on city streets.

Complete knowledge of the mind-as-brain might lead to new mind-altering drugs and electronic devices (hats with electrodes connecting the brain to a "happy machine"). We could use one software program to put our mind-brains into a business mode from nine to five and insert a "party disk" in order to program ourselves for weekend activities. A team of scientists using a "genius" program might bring science to complete knowledge of the physical universe. However, in order to function, these programs would have to be compatible with the enormous complexity and subtlety of our minds, whether purely physical in origin or otherwise. Our illusory, brain-produced minds would not disappear, turning us into mere robots. Technically such programs might make us smarter, but would they make us wiser? Is the wisdom we so desperately need to lessen human suffering and explore reality the product of mental agility alone? IBM's Blue Max, the computer that beat the reigning world chess champion, is smart, but would you want it for a friend?

Some neuroscientists may envision the practical applications of having a complete understanding of the brain the way geneticists view the human genome program. If we can know how each gene functions, we may be able to make alterations that can produce people free of genetically transmitted diseases and other genetic "defects." At such a time, all humans would essentially be born from test tubes. With a similar knowledge of the brain, we might attempt to eliminate unhappiness, stupidity, negative emotions—to create, in other words, a mental paradise— but who, then, would have the wisdom to decide which traits are defects and which phenomena contribute to happiness? Moreover, approaches such as these may merely represent an improvement on programs we already use, namely pills and alcohol. They allay symptoms but do not treat the causes of suffering. This is precisely the vision presented by Aldous Huxley in his famous novel *Brave New World,* which was written as a warning.

Would an understanding of how the brain produces the mind really clarify important metaphysical and philosophical issues? To say "Now we know! The mind and its core component, consciousness, arise from complex electrochemical processes" wouldn't necessarily give us new clues as to whether we as a species can know the realities of the universe, or how. Recall Immanuel Kant's critique of the limits of human knowledge. Proving scientifically that the "soul" is merely an electrochemical chip would be a blow to religion, but how would it help physicists or astronomers?

Knowing that our minds and senses are only physical mechanisms doesn't solve the paradoxes and complexities of the subatomic world. That's because the mind *as we use it individually,* not a physical organ, is still the final arbiter of what we consider true and not true. Whether "illusory," "nonexistent," "second-order representation," or "echo," our minds are nonetheless essential to our very identity. Even if the gray matter is the mind, only our first-person experience of mind can conclude that this or any other assertion is true. If we were to expose a human brain on the operating table and ask it, "Is the mind a product of you, the brain?" the gray matter itself wouldn't answer by forming itself into the shape of thumbs up or down. The entity that decides questions like "What is the mind?" will always be the mind itself, no matter what its origins, physical or otherwise. Likewise, the mind, not the brain, will decide whether it believes electricity to be a fluid or a flow of electrons, or whether gravity is best explained as a force of attraction or as the shape of curved space-time.

Suppose we used a modern computer as an example of a mind-brain. The "brain" could be compared to the chips, circuitry, and so forth that are unseen by the user, hidden within the chassis. The "mind" could be compared to the operating system, which interacts with the user. Even if the work of calculations is done by the computer's internal processor (its "brain"), and even if we view the programs facilitated by the operating system as "emergent properties" essentially identical with the rest of the hardware—without the operating system (the "mind"), the computer would be useless. Likewise even if the human mind, along with the individual personality associated with it, is illusory, it is this illusion with all of its intricacies, quirks and faults that decides all issues. It is the mind that becomes foolish or wise. If the brain appears and interacts with "oneself" as a mind, or even if the brain is "oneself," subjectively experienced mental states and processes still have to be accounted for. They do not vanish nor are they explained by simply equating them with brain functions. The mind, physical or not, still rules.

MIND AND BRAIN—A TWO-WAY STREET

We touched previously on research into the relationship between perception and the imagination. Studies of the brain and nervous system have revealed that in many cases the same neural systems and activities are involved in both perception and the imagination. For example, both high-level visual perception and the ability to manipulate imagery (imagination)—using memory, language, anticipation, and movement—share the same neural bases. Although the way this happens is subtle and complex and not perfectly understood, vision and imagination mobilize many of the same brain systems.

One important implication here is that our mental life, the weight of personal experience, influences the physical makeup and functioning of the brain. Recent discoveries of the brain's plasticity—its ability to be reconfigured by experience—support this conclusion. In a complementary fashion, brain systems, which are constantly being modified by experience, contribute to how we perceive. Thus, the qualitative, non-physical aspects of life experience are intimately connected with the physical brain. Whether the mind is illusory or not, it has a close, inter-active relationship with the gray matter, just as the screen and keyboard do with the computer's processor. Try as it might, neuroscience has failed to reduce the mind to a purely biological mechanism.

By Pushing the Envelope

The general conclusion, as one research paper puts it, is that "percep-tion . . . can be seen as constrained imagination."[2] If that is so, then the big question for philosophers from Plato to Kant and scientists from Galileo up to the present is, Constrained by what? What is it "out there" that is interacting with our senses to limit our perception, rather than permitting the imagination to present us exclusively with hallucina-tions? In its four-hundred-year history science has been unable to satis-factorily answer that fundamental question, originally stated as, What is the relationship between the human mind and the universe? We have suggested it was doomed from the start, having begun its journey based on incorrect assumptions. In doing so science has exhausted numerous promising theories, always optimistic that the light at the end of the tun-nel is near. Yet if there is some underlying reality to the universe, then when we seem to have found it, why won't it hold still?

This conundrum is also a product of the zeal and thoroughness of sci-ence, but it may offer the hint of a solution. By pushing the envelope of knowledge so diligently science has come full circle from ignoring the mind completely to hitting it full-on. The mind seems to be the key to our most sought-after knowledge, both as an impediment when ignored and as a doorway when included.

Let's consider from a new angle the relationship between perception and imagination in light of recent neuroscientific research. According to this view, the difference between pure imagination, such as the fantasies of dreams and daydreams, and perception is that the latter is constrained by input from the outer environment. If I perceive a rose and see its color as "red," the appearance of that color to my consciousness is assumed to be a response to an outer stimulus. That perception is conditioned by my eye and brain. Color-blind persons, and most likely other organisms with visual senses, do not see the color red as most humans do. "Red," which is a human concept/experience, doesn't exist for them as it does for us. Their differing visual sense organs respond to the original stimulus differently.[3] So we are left again with the "universe," "environment," or "phenomena"—some "real" cause of "red"—existing in some manner we cannot perceive directly. Apparently our only access to the "real world" is through scientific laws, often stated as mathematical formulas, that somehow stand for, represent, or symbolize the unperceivable true reality. However, here we Westerners really haven't come any closer than Plato, who 2,400 years ago proposed a hidden world of "ideal forms" as the underlying reality of the universe. Yet we continue to ask the same question over and over: Our senses and imagination are constrained by *what*? Perhaps, finally, we would do better by changing the question and asking, Constrained *how*?

A MIDDLE WAY?

Experiments in relativity and quantum physics, such as those demonstrating the wave/particle characteristics of certain phenomena, have a process orientation. They reveal that looking for ultimate answers in the form of absolute physical entities and states gives only a partial view of reality. This would be like trying to appreciate the game of basketball while only watching the basket. In addition they strongly suggest that the mind of the observer is an active participant in what were once considered isolated, purely physical realities. Scientists (in their white coats, symbolizing the clean, objective separation between them and their experiments) release photons into their apparatus, place a detection device that will reveal particles, and get particles. If they place a detection device to reveal waves, somehow the photons become waves. Yet is there a wave/particle "photon-thing" out there to be detected, or is the experiment provoking or creating a process and nothing more?

A similar process-oriented view of the mind as a fundamental participant at all levels of reality is found in many contemplative traditions, including Taoism, Buddhism, Hinduism, Islamic Sufism, and Jewish mysticism.[4] One of the most cogent presentations of this hypothesis is found in the *Prasangika Madhyamaka* philosophy of Buddhism.[5] Here we will concentrate on Buddhist views and analyses because we are most familiar with them and also because they have had a long and relatively undisturbed history of systematically exploring the relationship between the mind and phenomena—2,500 years and running.

Madhyamaka means and is equivalent to the Buddhist philosophy called "the "Middle Way." It arrives at reality by avoiding extremes. One extreme is called "absolutism." This is likened to a view stating that phenomena have absolute reality: they exist independently of one another, have inherent existence, have their own individual essences, and are "really real," independent of any mode of inquiry or conceptual framework. The other extreme is "nihilism," which utterly denies any reality, declaring that all appearances are hallucinations with no real basis. The Middle Way is closely linked with the theme of interdependence, which is to say that reality is a cooperative process that depends upon all participating phenomena. A game of professional basketball arises in dependence on its players, the basketball court, the rules of the game, the arena, the fans in the stands, the ticket sellers, and so on. Take away any one of these and there is no game of professional basketball as presently defined.

As we explore the Middle Way more deeply, be aware that its overall position is opposed to the commonsense realism we are accustomed to and that is the foundation of scientific realism and scientific materialism.

STEPPING-STONES TO EMPTINESS

There are three Buddhist philosophies that are often used as stepping-stones to an understanding of the Middle Way, and they are similar to many views we hold in the West. The first, *Vaibhashika*, considers tiny, irreducible atoms to be the only physical phenomena having absolute existence. It also believes that brief, irreducible moments of consciousness are the absolute level of the mind. So Vaibhashika is an atomistic philosophy—everything reduces down to some ultimately small, but absolutely existing basis. Physics, as we saw, has been guided by similar beliefs. This is a form of physical, and mental, reductionism.

The Middle Way rejects these ideas by pointing out that for something to be irreducible—the smallest thing in existence—it must be absolutely isolated from everything else and therefore cannot influence or be influenced by anything. For example, if the ultimate constituents of physical reality were irreducibly small particles or atoms, how could they combine to form larger things? If atom A were attached to atom B to form a larger particle, then some side of A (say the left) would have to touch some part (the right) of atom B. In that case, of course, the atoms would have to have parts ("left" and "right") and could not therefore be the smallest components of matter, because the parts would be smaller. Similarly, if one calls the parts or particles of an atom the ultimate constituents of matter, the same problem occurs when trying to construct larger phenomena from them. The dilemma repeats itself ad infinitum. The Middle Way therefore concludes that independently real atoms that participate in causal interactions cannot exist even in principle.

In like manner, if human consciousness were composed of ultimately indivisible, short moments of consciousness, how would they ever combine to form a continuity of consciousness? How could we remember anything if moments of consciousness were isolated from one another? For us to have memories, they would have to join in a temporal sequence, but if they joined, it would destroy their identity as isolated moments. In this way, a provisional Vaibhashika position is refuted by a definitive one from the Middle Way: matter and consciousness are found to be "empty" of independent, permanent, inherent "atomic" constituents.[6]

Is this conclusion different in principle from what the analysis of quantum physics has done to the atom and its subatomic constituents? The conventional scientific definition of a truly existing particle requires the simultaneous measurement of its position and momentum. It has to be "somewhere" moving "somehow." However, as we saw in part 1, precise position and momentum not only cannot be measured but do not even exist simultaneously in a subatomic particle. Furthermore we saw that we cannot even be sure that the concept "particle" has any meaning independent of our conceptual frameworks. Depending upon what kind of experiment or measuring device we use, we may detect either particles or waves. As we have seen, quantum physics seriously questions the absolute, independent, inherent, permanent existence of subatomic entities. Relativity, too, questioned concepts that were assumed to be true in classical physics, fundamental notions such as absolute space, time, and mass existing independently of the motion of other phenomena. Yet, like the Middle Way, physics can still deal with them as *relatively* existent. By doing so the atomic bomb was invented, and that certainly exists at some level!

LABELS—DON'T BELIEVE EVERYTHING YOU READ

A second Buddhist view of reality, *Sautrantika*, claims that both physical and mental phenomena are "concrete," that is, they are real and exist independently of any conceptual framework, though they are constantly changing and engaging one another. These are ultimate truths in this system. However, the Middle Way would say that there is a contradiction between something's being both inherently real and its having an interdependent relationship with other conditioned phenomena. If an object were absolutely real, intrinsically bearing its own parts and attributes, how could it possibly be dependent on other phenomena, and if it had an absolute nature of its own, how could that possibly be influenced by or interact with other phenomena?

This argument applies as well to the conscious mind: If mental phenomena are constantly arising in dependence upon the brain, environmental influences, and earlier mental processes, how can any of them have an essential, absolute nature of their own? Independent, substantial entities, as proposed by Sautrantika, are found nowhere in nature. Physical and mental events are merely labeled in various ways, and there is nothing absolute about applying labels to things. We tend to think of objects such as a wooden table as being concrete, stable, and absolutely real. Yet a table does change over time: it may become chipped or broken, and it may be sanded and repainted. When that happens, it's no longer the same table. In fact, one could just as easily view this "table" as merely a stage in the history of some wood, or of some trees, or of the soil, rain, and sunlight that contributed to the trees' growth. In the same way that we apply the label "table" to this object, we call things either "mind" or "matter" due to a set of conventional labels. However, both are empty of absolute existence.

Here the Middle Way makes a key distinction between causes and regular occurrences or patterns. Sautrantika and other philosophies East and West believe in causes—that a cause has within it a power to affect something. This power is something inherent within the cause and gives the cause itself (and any chain of causes and effects) an absolute, inherent existence. The Middle Way, by contrast, finds no more than conventional regularities: when x occurs, y tends to happen also. X is not considered the absolute "cause" of y. No inherent "power" to cause something is implied. Mental and physical phenomena are constantly changing and therefore not absolutely real. Neither is there any real, concrete quality behind or within those changes that gives causes any reality either, that bridges one thing in a sequence to another. The smell of food does not inherently "cause" hunger. They simply occur on occasion, one after the other. The oxygen in water does not inherently "cause" iron to rust. This is just an observed regularity. Hunger, rust, and all other phenomena exist within complex, interactive webs that exhibit certain regularities. Notice that this is a position similar to the instrumentalist, pragmatic, and antirealist philosophies we saw earlier.

LOSING OUR MINDS

If Vaibhashika and Sautrantika lean toward attributing absolute reality to physical and mental phenomena, a third Buddhist philosophy, *Chittamatra,* states that mind and only mind has absolute existence. Here the mind is regarded as truly existing, whereas external objects consist only of appearances having no reality distinct from the mind. The images from the outside world that appear to consciousness do not exist apart from the mind that apprehends them. However, the Middle Way denies the absolute existence of the mind. Thoughts, images, feelings, and other mental phenomena do appear, but "the mind," as an independently existent subject, does not appear. "Mind" is merely a label we apply to a collection of experiences just as "professional basketball" is a label applied to the collection of phenomena mentioned earlier.

In its analysis of the positions of all three philosophies—Vaibhashika, Sautrantika, and Chittamatra—the Middle Way avoids attributing absolute existence to either physical or mental phenomena. Both exist conventionally but are empty of absolute existence. Divisions are constructed by the conceptual mind and have no reality apart from these labels. When it comes to physical phenomena, quantum physics and relativity have severely questioned the absolute existence of space, time, and matter. Regarding the mind, however, neuroscience sometimes questions even its conventional existence, regarding it as a mere apparition with no causal power of its own. This view of mind would be similar to the Chittamatra view of matter. At best neuroscience accepts mind as a conventional reality, but one that "emerges" somehow from the absolutely existing physical brain. We'll examine this in more detail later.

According to the Middle Way, the absolute, underlying truths sought by realists do not exist, and the search for them is as futile as seeking the pot of gold at the end of a rainbow. Where does that leave us? With "conventional truths." The Middle Way defines a conventional or relative truth according to three criteria: (1) it is accepted as valid within a given community, (2) it doesn't contradict some other conventional truth accepted within that community, and (3) it doesn't contradict what Buddhism terms the "ultimate" level of truth—the truth that all phenomena are empty of inherent existence. (The Buddhist term *ultimate truth* has nothing to do with the underlying mechanisms sought by scientific realists. It refers to emptiness, a topic we shall explore in some detail below.) A "conventionally accepted truth" could refer to something as broad as a scientific law (which might be contradicted in the future), down to, say, the truths about what makes a wine superior (as accepted by a small group such as professional wine tasters). So something is conventionally true within a specific context. It does not require, and in fact cannot have a foundation in, some underlying, absolute reality. Yet while lacking absolute reality, things can still function perfectly well.

A Diamond Mind

An example: Within a given system of conventional truth, for instance our everyday way of looking at things, the diamond is the hardest natural material in existence. Since we know of no harder material in nature, there is no conventional truth contradicting this, though we are willing to accept that a substance harder than a diamond might be found. In everyday, commonsense terms the diamond is the hardest natural material known. (Indeed, there are now some artificially made materials that are harder than the diamond.) We accept that the diamond's hardness is not absolute and that its status is dependent on criteria such as our having heard that scientists "say" the diamond is the hardest material, or perhaps our having been unable to break a diamond with a hammer, our not having heard contradictory information, and so on. "The diamond's hardness" exists, therefore, within a web of knowledge and observations but is not absolutely, independently, inherently true. In Middle Way terms, it is a conventional truth.

It is our commonsense, realist notions that introduce absolutes. For instance, they prod us into believing that diamonds not only are absolutely hard but that they absolutely exist, just as they tell us that all physical objects have absolute existence. We believe that a diamond, which we can see and touch, has some essence that makes it a diamond and that that essence makes it independent of all other kinds of physical objects. As a diamond, therefore, it is absolutely real as well as absolutely hard. This the Middle Way denies. Like the red color of a rose, the diamond that we perceive depends on our senses and our minds for its existence. Our minds and senses are conditioned, in turn, by our human physiology as well as by our individual cultural, educational, and personal backgrounds. All these together create the idea of "diamond-ness." Without them we have no way of knowing whether we could even perceive a diamond or not. To illustrate, imagine that we held a diamond in the air as a bat passed overhead. Perhaps the bat would perceive something in the space between our fingers. But would it be a diamond?

If we analyze the diamond to determine whether it has absolute existence, we discover that diamonds don't exist independently. Diamonds are produced when carbon is put under enormous pressure over a long period of time. If the diamond is produced by conditions over time, it can also be destroyed over time. The diamond could even be viewed as a moment in the life of some carbon—thus, diamonds are not "forever." A diamond might have once been graphite, a carbon substance that happens to be one of the *softest* of materials. We know that diamonds are also made up of smaller components—atoms of carbon in a specific arrangement—and furthermore that these atoms are composed of sub-atomic particles held together by energy. Yet any competent physicist will tell you that we don't really know what energy and subatomic particles really are.[7] We have no certain knowledge of their absolute existence. Under analysis the diamond falls apart.

The diamond, then, fulfills the Middle Way's three criteria for a conventional truth, but not our commonsense, realist notions of absolute realities: (1) it is conventionally accepted that the diamond is the hardest natural mineral; (2) this does not contradict some other conventional truth; and (3) the diamond's hardness and its existence are not absolutes. The diamond that we perceive exists only in relationship to its attributes (clear, hard, heavy, and so on), its cooperative conditions (carbon, heat, and pressure, time, discovery by humans, and so on), and our naming such things diamonds. The diamond emerges as conventionally existent within a web of conditions; it is part of an interdependent process in which the mind plays a crucial role.

Let's look more deeply into the mind's role here. We have seen that the notion "seeing is believing" is very limited. Our senses do not provide an open window on reality. They are specifically designed for human bodies, and their use has been conditioned both by human history and by our brain and nervous system (themselves modified by human history and each individual's experience). Our mind that decides something is "known" or "understood" might be viewed as a composite that includes the senses, brain and nervous system, memories, experience, education, and so forth. Even when making relatively simple observations, when the mind says, "Ah-ha! I know!" it is based on the weight of habit— human, cultural, and individual. Every common act of knowing is heavily conditioned.

Therefore, diamonds and all other phenomena are "empty" of the conditioned notion of absolute existence. This brings us to a key term— *emptiness*. This word has often been translated, erroneously, as "nothingness," which has led to all kinds of mistaken views about Buddhist philosophy. "Nothingness," which would deny the existence of ordinary entities, such as diamonds, is one of the nihilistic positions that the Middle Way avoids. The reason is that diamonds, rainbows, opinions, memories, and so on do have conventional existence, and are not "merely nothing."

Conversely, there is sometimes a temptation to view emptiness itself as an absolute, as an underlying "voidness." Clearly, however, that cannot be true because emptiness is always empty *of something*. It has a relationship of dependence upon some phenomena that can be characterized as empty of absolute existence. Emptiness is merely the manner in which phenomena exist. Therefore "emptiness" is itself empty.

THE VALUE OF RAINBOWS

Rainbows play a special role in the lore and literature of Tibetan Buddhism in particular. Their appearance in the sky is considered auspicious. They are often seen in abundance when great spiritual events or accomplishments take place. It is said that one of the greatest achievements of a Tibetan yogi is to turn his body into rainbow light at death. Moreover, rainbows are a perfect example of interdependence, of the Middle Way.

Are rainbows real? A rainbow appears only when the following conditions are met: there is moisture in the air, there is sunlight, and there is an observer at the proper angle to the moisture and sunlight who is paying attention and who can recognize a rainbow. But where is the rainbow located? Is it in the moisture? In the sunlight? Somewhere in between? Between these and the observer? On the retina of the observer? Or does it exist solely in the mind of the beholder? Since the rainbow is a dependently related event, it has no absolute location any more than it has an absolute existence. We don't know if other creatures can perceive rainbows as we do (though it may be that they perceive phenomena perhaps as amazing and beautiful as rainbows that we are completely unaware of).

According to the Middle Way, everything we perceive, including our thoughts and even ourselves, are as illusory as rainbows. They appear to our minds and senses and therefore "exist" (if we denied their existence, that would be a nihilistic position). But their existence is not permanent or absolute. When analyzed they are seen to have no essence; they do not exist inherently; and they are not independent from their context (which would be the absolutist position).

Quantum physics was not alone in our examples from science that parallel this view. Recall how the Newtonian notions of absolute space and time fell apart under the analysis of relativity. If the speedometer of our car says we are moving at the rate of 70 mph down the highway, we would normally accept this as an absolute truth. However, as we saw earlier, depending on one's latitude, the earth is moving at anything from zero to 1,000 mph as it rotates. In like manner, the earth is rapidly moving around the sun, the sun takes about 200 million years to travel around the Milky Way, and every object in the universe is moving in relation to every other one. Thus, how fast are we traveling? There is no absolute answer, only a conventional answer given within a conventional, relative context.

Therefore, both physics and the Middle Way pull the rug out from under our commonsense notions of reality, based as they are on things having absolute existence. At each step in the analysis of a supposedly "real" object, a piece of the rug disappears. The realist habit then seeks a "deeper or more fundamental reality." However, each fallback position is equally insecure, because there is no rug supporting any realist position. The absolute realities being sought won't hold still. This is precisely the dance that played out over a thirty-year period, originating in Einstein's realist-locality assumptions in the EPR paradox (in which Einstein and his colleagues assumed the absolute reality and limited, local effects of a supposedly closed system) and ending in its disproof by Bell's inequalities (showing that such systems are open—that is, they are entangled with wider contexts and indirectly include the mind that measures them), which we examined earlier. Given where science has taken us, this story may even be an apt metaphor for the history of science itself.

The mind, then, is necessarily at the heart of every assertion of reality. Spiritual traditions have discovered this through contemplation and occasionally from instantaneous insights and revelations. Science has begun to suspect this, because ignoring the mind has backfired. Could the two fruitfully collaborate? To answer that, scientists must ask, Do contemplative traditions have the kind of rigorous, organized, empirical knowledge of the mind that would permit cross-fertilization with science? That is to say, is there a common ground for the cooperative development of a science of consciousness?

Deciding What's True 7
Objectivity and Empiricism in Contemplation

I N THE MOVIE *2001* (based on the novel by Arthur C. Clarke), a black, metallic monolith is discovered by a group of prehistoric apes. Placed in their midst by some mysterious, advanced civilization, this object has the power to make the monkeys suddenly smarter. Soon they discover the first tool: a thighbone that can be wielded as a weapon. The monolith, it is suggested, is responsible for these apes' eventual evolution into a more intelligent species—human beings. Were we to seek the origins of contemplation among humans, we could imagine any number of possible scenarios, even something as bizarre as a black monolith. Several spiritual traditions state that the source of their contemplative practices is extraterrestrial, at least in the sense that divine or enlightened beings dwelling in other dimensions brought them to earth. Whether one takes this as legend or fact, pinning down the origins of contemplation is largely guesswork.

Contemplation reflects an interest in the mind's interior realm. Since few in the outward-facing modern world are attracted in this direction, it is especially hard for us to imagine how contemplation might appear and then evolve into an important focus in many spiritual traditions. We hear of Moses coming down from the mountain with the Law, and Muhammad's requirement that Muslims pray five times a day, but, at least in the West, we don't often hear of prophets declaring, "Thou shalt meditate!" Even if gods, buddhas, or other special beings did somehow spark an interest in contemplation among *Homo sapiens,* it is likely that contemplation actually developed among us as an oral tradition long before anything about it was written down. However, since we must rely on written records and can only guess at the meanings suggested in prehistoric symbols, we don't know with certainty when or where the practice of contemplation began or what early contemplatives were up to.

A BRIEF HISTORY OF CONTEMPLATION

Contemplation (from the Latin *contemplari,* "to gaze attentively, observe") is an act of engaged attention. The word *meditation* is sometimes used as a synonym. Whether the object of this attention is external or internal, physical or mental, the mind is concentrated in such a way as to eliminate distractions and bring the chosen object into clearer focus. Beyond this very general definition, the methods and aims of contemplation vary greatly. An art student could be said to contemplate the beauty of the Mona Lisa, whereas some forms of Taoist meditation have as a goal immortality itself. One could also stretch the definition of contemplation to include devotional rituals such as those found in the bhakti tradition of Hinduism exemplified by the repetition of a song or the mantras in Hare Krishna rituals. Here, too, the immediate aim is to concentrate—in this case on devotion—thereby eliminating distractions.

Our knowledge of contemplation in the West begins with mystical brotherhoods of the eastern Mediterranean such as the Pythagoreans and the Essenes.[1] Pythagoras (c. 570–c. 495 B.C.E.), best known today for his contributions to mathematics, founded a contemplative community in southern Italy that combined spirituality and science. He claimed direct knowledge of reincarnation, according to which the soul is immortal and is reborn in both human and animal incarnations. Legend has it that he arrived at that belief from having remembered some twenty of his past lives as well as the previous lives of others. It was also said that Pythagoras could work miracles. His school has been described as devoted to the purification of the body and the mind, but we have no records of the specific contemplative methods employed there. It is interesting though, and relevant to our purpose here, that the Pythagorean Society became the first scientific school of Greece.

The Essenes, a small Jewish monastic community in Palestine, date from the second century B.C.E. We know they were ascetic in nature, but we have no records as to whether they practiced some form of contemplation or what those practices might have been. There have been suggestions that John the Baptist and Jesus Christ were members of the Essene Brotherhood. The Neoplatonists came a bit later—about the third century C.E. They believed that contemplation and purification led to union with "The One," or "The Good." However, this was apparently considered a fleeting mystical state. The school's leader, Plotinus, was said to have achieved it only seven times. Here again we have no clear detailed knowledge of the practices they employed.

Contemplation as practiced by Christian mystics appeared very early in the history of the church. Although a form of mysticism is attributed to Jesus's disciples John and Paul, specific information on meditative practices doesn't appear for several centuries. By the Middle Ages this introspective exploration of the mind for spiritual purposes went on quietly behind the scenes, running parallel to the official church doctrines emphasizing faith and worship rather than exploration of the mind.

A PATHWAY TO GOD

The writings of Christian contemplatives, whose goal was direct experience of the divine, show that they saw the mind as the pathway, though a tricky one, to union with God. By concentrating one's mind on a single word (*God* or *love*), Saint John of the Cross (1542–91) advised that "thou shall smite down all manner of thought under the cloud of forgetting" that would lead one closer to God. In a similar vein, John Cassian (fifth century), quoting Abbot Isaac, wrote, "Take a short verse of psalm, and it shall be a shield and buckler to you against all your foes." "The aim of every monk, and the perfection of his heart," said Isaac, "tends to continual and unbroken perseverance in prayer, and, as far as it is allowed to human frailty, strives to acquire an immovable tranquility of mind and a perpetual purity . . . ." Indeed, many of the statements gathered by Cassian in his book, *Conferences,* comprise a manual of contemplation as practiced by fifth-century Christian "desert fathers" in remote regions of Egypt and Syria. Thus prayer, usually viewed as an emotional or intellectual act, could be for Christian mystics a means for calming and clearing the mind. For them, a mind emptied of ordinary thinking was capable of hearing the word of God: "No one can attain this birth (of God realized in the soul) unless he can withdraw his mind entirely from things," said Meister Eckhart (1260–1328).

Saint Thomas Aquinas analyzed the practice of contemplation as "hindered both by the impetuosity of the passions and by outward disturbances," and explained that moral virtues were their antidote. Saint Francis de Sales gave this advice on concentration: "If the heart wanders or is distracted, bring it back to the point quite gently and replace it tenderly in its Master's presence." Beyond mere concentration, Saint John of the Cross spoke of the need to "look within" in order to "transcend the human mind." Such statements, drawn from an extensive literature of Catholic mysticism, suggest an analysis of the mind based on direct experience, on introspection. Teachers of contemplation had to guide their pupils with advice based on firsthand knowledge of the mind, which may have been quite extensive.

In many cases this path to union with God may have been guided by an overall philosophy, which has been broadly outlined in three stages by the modern Belgian Jesuit philosopher Joseph Maréchal: In the first stage, through prayer and asceticism the ego is gradually integrated through full acceptance of the idea of a personal God. In the following stage, the soul experiences an ecstatic contact or union with God, a transcendent revelation wherein the normal physical faculties are often suspended. A final stage of maturity is described as "a kind of readjustment of the soul's faculties" by which it reenters social contact "under the immediate and perceptible influence of God present and acting in the soul."[2]

A long history of contemplation is found in the Eastern Orthodox Church. As early as the fourteenth century, the Hesychasts, monastics living on Mount Athos in Greece, were known to prepare for the "Jesus Prayer" (lengthy repetition of the name of Jesus) with breathing practices and concentration on internal organs. The writer Aldous Huxley commented on such techniques, saying, "[T]he constant repetition of 'this word GOD or the word LOVE' may, in favorable circumstances, have a profound effect on the subconscious, inducing the selfless one-pointedness of will and thought and feeling, without which the unitive knowledge of God is impossible."[3]

134 CONSCIOUSNESS—COMPLETING THE PICTURE

Although contemplation is relatively rare in Protestantism, similar goals and methods are found occasionally. The eighteenth-century English theologian William Law said, "[O]ur own silence must be a great part of our preparation for . . . hearing what the Spirit and voice of God speaketh within us . . . ." This attitude is reflected in the only modern Protestant denomination that emphasizes meditation—the Society of Friends, or Quakers. Here meditation is conducted in a group setting. The parishioners "wait on God's word" in communal silence until someone is moved to speak. Contemplation, often in the form of extended prayer, still plays an important part today in several Roman Catholic orders, among them the Trappists and Carmelites (the latter exemplified by the Missionaries of Charity founded by Mother Teresa).

It is problematic to extrapolate precise meditative procedures from the scant information available on early Christian and eastern Mediterranean contemplative communities. In many cases the techniques were secret. Often the experiences resulting from these practices were described as short-lived. Although it can be assumed that in some cases, especially in monastic settings, some form of peer review existed as part of the transmission of contemplative knowledge from teacher to student, we don't know the precise nature of this either. Finally, many of these sects and mystical brotherhoods had short or tumultuous histories, which may not have allowed for an effective methodology to develop, establish itself, and mature.

MAPS IN GREATER DETAIL

The picture begins to clear somewhat as we move East. Sufism, which emerged from Islam in the late tenth and early eleventh centuries, also uses word repetition, namely the practice of *dhikr*, "reminding oneself," where words including the name of "Allah" are repeated either silently or aloud, sometimes in coordination with breathing. However, depending on which particular Sufi brotherhood one belongs to, there are other contemplative practices, ranging from the famous "whirling" dances of the dervishes to techniques of seated meditation. In Sufism we find a rich literature, developed over nearly a millennium, along with a tradition of master and disciple. Students are expected to purify themselves of selfishness through introspection and inner struggle to the point that they attain *ikhlas*, absolute purity of intention and act. Spiritual masters (*shaykh* or *pir*) lead their disciples through an ascending sequence of spiritual steps (*maqam*). Disciples are to place themselves unreservedly in the hands of their master "like a corpse in the hands of the washer."

Islam is a religion of surrender to the will of Allah, and Sufism reflects this in its emphasis on love and devotion toward this supreme deity. Although its literature includes philosophy and cosmology, Sufism places emphasis on devotional poetry. Wisdom, spiritual symbolism, devotion, and ecstatic descriptions are combined in the poems of Jalal e-Din Rumi (1207–73), the Sufi poet best known in the West. All of these qualities play an important part in the Sufi path to union with Allah. However, at least for the uninitiated, they do not present a clear explanation of the methods of contemplation used to achieve the exalted states referred to. This information is gradually becoming available in the West through the translations of treatises on Sufi meditative techniques and presentations by modern Sufi masters, some of whom have had an active following in the West.

We also have access to meditative practices in handbooks from the Taoist tradition of China. The origins of Taoism date from the fourth century B.C.E. with Lao-Tzu, the author of the *Tao Te Ching* (*The Classic Way of Power*), considered its principal founder. Over its long history, Taoist contemplation has been cultivated by both solitary recluses and monks living in monasteries. One of its principal goals is *wu-wei*, "inaction" or "nondoing." In reference to this achievement Lao-Tzu said, "Having attained perfect emptiness, holding fast to stillness, I watch the return of the ever active Ten Thousand Beings [symbolizing totality]." In texts containing discourses between masters and disciples, questions are introduced by students based on their meditative experiences. Using this feedback, these masters carefully guide their students toward the goal.

The *T'ai I Chin Hua Tsung Chih* (*Secret of the Golden Flower*), a Taoist meditation guide written in 1794, contains detailed instructions for an alchemical process to increase longevity and attain spiritual realization.[4] Precise illustrations demonstrate a process of "circulating the light" that combines visualization with breathing practices. There are also illustrations alluding to different stages of spiritual development. Among the chapter headings are "Circulation of the Light and Making the Breathing Rhythmical," "Mistakes During the Circulation of the Light," and "Confirmatory Experiences During the Circulation of the Light." However, Carl Jung, who wrote a preface to the German and English translations of this work, warned that the alien language and concepts would not permit a Westerner enough understanding to even begin to make practical use of this handbook. Like many other such guides, this text was probably used in conjunction with oral instructions, without which it would be useless in terms of a detailed understanding of Taoist contemplative methods. Indeed, the interpretations given by the translator, Richard Wilhelm, abound with concepts reflecting the European psychoanalytical school of psychology. These may obscure rather than illuminate the original nature of the text.

Despite their vagueness, the hints provided in Christian mysticism, sects from the eastern Mediterranean, Sufism, and Taoism do give us some idea of the goals and methods of contemplation. These are confirmed, expanded, and greatly clarified when we examine the great contemplative traditions that emerged from India and later spread to Southeast Asia, Tibet, China, Mongolia, Central Asia, and Japan.

MOTHER INDIA

Hinduism traces its source to the Indo-European people who settled in India in the second millennium B.C.E. It is based in the Vedas, texts whose date and origin are unknown, but that, as early as the fifth century B.C.E., were held to be expressions of eternal truths. The term *Hindu* covers a wide range of spiritual groups and practices including six traditional philosophical schools and innumerable subsects. A wide range of contemplative practices has developed in the majority of these schools. They are nested within rich traditions that include canonical texts (such as the Vedas, and Upanishads), cosmology, ethics, scholarship, commentaries, ritual, monasticism, meditation guides and handbooks, and the master-disciple relationship. An immense literature, then, is available to translators and scholars, thus enabling Westerners to gain an understanding of many aspects of Hinduism and its contemplative procedures.[5]

Ideally such intense activity might lead to a system of knowledge based on empirical data of a first- and second-person nature, as found in the master-disciple relationship, reflected in and supported by a critical literature. Based on an oral tradition where specific and intimate knowledge of contemplation is passed on individually or to groups in monasteries and lay retreats, and backed by written accounts by scholars and sages, masters carefully monitor their students' progress, reflected in first-person reports of meditative experience. If this process consistently leads to bona fide experiences and forms of realization, it can thus be said to be based upon empirical data, even though subjective in origin. This is a crucial point. As we shall see below, this carefully monitored process within the master-disciple relationship has parallels in science, which can provide grounds for a dialogue between contemplatives and scientists.

YOGA AND YOGIS

Teachings from the Yoga school of Hinduism provide a detailed illustration of such contemplative procedures. Yoga (literally, "union") is based on the *Yoga Sutras* of Patañjali (apparently written between the second century B.C.E. and the fourth century C.E.). It includes not only the familiar hatha yoga physical postures and breathing exercises but a complete training whose ultimate aim is the union of the individual soul or consciousness with the universal (Atman, Brahman, or God). According to the yoga teacher B. K. S. Iyengar, yoga training entails "the coordinated functioning of the yogin's body, senses, mind, reason, and self."[6] The techniques and attitudes involved include concentration and meditation, the aforementioned master-disciple relationship, mantra (repetition of sacred syllables, words, or prayers), proper motivation, perseverance, zeal, devotion, study, and ethical considerations (including nonviolence, truth in conduct, celibacy, not stealing or coveting, not hoarding possessions, contentment, and purity and cleanliness of the body). Clearly this describes a concentrated, highly involved undertaking.

The *Yoga Sutras* provide a window into the details of yoga's view of the mind and its possibilities. They are divided into four main chapters: "Integration" (with 51 aphorisms), "The Path to Realization" (55), "The Extraordinary Powers" (56), and "Freedom" (34). The short sampling of aphorisms listed below was chosen merely to hint at the diversity and

depth of subject matter and the detailed pedagogy of yoga's mind training. Keep in mind, however, that even the entire collection of over 180
aphorisms comprises only a rough guideline within a process requiring
the oral teachings of an experienced meditation master. The subtlety of
insight into the mind's functioning and possibilities, as the Yoga tradition conceives them, can only be imagined from these examples.

From Chapter 1: "Integration"

(1.1) Now, the teachings of yoga.

(1.2) Yoga is to still the patterning of consciousness.

(1.12) Both practice and nonreaction are required to still the
patterning of consciousness.

(1.14) The practice becomes fully rooted when it is cultivated
skillfully and continuously for a long time.

(1.24) *Ishvara* [or "God"] is a distinct, incorruptible form of
pure awareness, utterly independent of cause and effect and
lacking any store of latent impressions.

(1.30) Sickness, apathy, doubt, carelessness, laziness, sexual
indulgence, delusion, lack of progress, and inconstancy are
all distractions that, by stirring up consciousness, act as bar-
riers to stillness.

(1.33) Consciousness settles as one radiates friendliness, compassion, delight, and equanimity toward all things, whether
pleasant or painful, good or bad.

(1.41) As the patterning of consciousness subsides, a transparent
way of seeing, called coalescence, saturates consciousness; like
a jewel, it reflects equally whatever lies before it—whether subject, object, or act of perceiving.

(1.47) In the lucidity of a coalesced, reaction-free contemplation, the nature of the self becomes clear.

From Chapter 3: "The Extraordinary Powers"

(3.1) Concentration locks consciousness on a single area.

(3.2) In meditative absorption, the entire perceptual flow is
aligned with the object.

(3.12) In other words, consciousness is transformed toward focus as continuity develops between arising and subsiding perceptions.

(3.18) Directly observing latent impressions with perfect discipline yields insight into previous births.

From Chapter 4: "Freedom"

(4.4) Feeling like a self is the frame that orients consciousness toward individuation.

(4.5) A succession of consciousnesses, generating a vast array of distinctive perceptions, appears to consolidate into one individual consciousness.

(4.15) People perceive the same object differently, as each person's perception follows a separate path from another's.

(4.16) But the object is not dependent on either of those perceptions; if it were, what would happen to it when nobody was looking?

(4.20) Furthermore, consciousness and its object cannot be perceived at once.

(4.34) [Last aphorism:] Freedom is at hand when the fundamental qualities of nature, each of their transformations witnessed at the moment of its inception, are recognized as irrelevant to pure awareness; it stands alone, grounded in its very nature, the power of pure seeing. That is all.[7]

Without going into any of the implications of these selected aphorisms, we do get the impression that Yoga has made both a deep, long-term study of the mind and developed a subtle, detailed method of imparting this knowledge to students. The methodology is empirical and based on peer review. The meditation master is expected to have achieved a high degree of realization before becoming a teacher. The student is guided through a lengthy and progressively subtler sequence of practices leading to specific experiences personally monitored by the teacher. Moreover, the entire process is backed not only by the literature of the Yoga philosophical school but by numerous similar treatises (and meditation practices) derived over several millennia from the Hindu religion as a whole. This body of knowledge has also been subject to the influences of and challenges from other contemplative traditions of the Indian subcontinent including Buddhism. Furthermore, public, scholarly debate over the truth and efficacy of philosophical and contemplative systems has been a common feature of Indian history.

Additional insight into the long history of the Indian sciences of mind can be gleaned from the reports of Siddhartha Gautama, a prince of the Sakya clan in what is now southern Nepal. In the sixth century B.C.E. Gautama, later known as the Buddha, became a wandering ascetic seeking out the most able meditation masters of his time. His first teacher, Alara Kalama, who headed a monastic community, taught Gautama "the base consisting of nothingness." Gautama quickly mastered this inner knowledge to such an extent that Kalama invited him to share leadership of his community. However, Gautama was seeking dispassion, the fading of lust, cessation, peace, direct knowledge, and enlightenment, and therefore moved on to a second meditation master, Uddaka Ramaputta. From him Gautama mastered another meditative state known as "neither perception nor nonperception." Again dissatisfied, Gautama practiced extreme asceticism for six years before finding a middle path between the harshness of self-mortification and the indulgence of the senses, leading him finally to enlightenment. The implication here is that as early as the sixth century B.C.E., one could find in India a variety of teaching traditions systematically expounding knowledge of the mind and consciousness.

But Is This Science?

Does any of this mean that ancient India developed a viable science of the mind? Modern science would have at least three major problems with this suggestion. First is the taboo against subjectivity. Rather than using surgery, electrodes, or magnetic resonance imaging devices, Hindu yogis explore the mind from the subjective, first-person perspective and monitor their discoveries in detailed reports to a mentor. Second, students of yoga are guided to an expected, preconceived set of experiences based on the authority of previous "realized" yogis or saints. The method of development—meditation—may be exploratory, but specific results are expected: in the Hindu example, progressively more subtle states and experiences leading ultimately to union with God. For both yoga and Buddhism this points out important differences with scientific attitudes toward authority. Traditional Buddhists rely deeply on the authenticity of the Buddha's enlightenment and the insights that came with it, though they may certainly question their own understanding of those insights. Scientists, on the other hand, regard no individual scientist or scientific theory as being ultimately authoritative.

Science views experimentation as leading to *new* knowledge, which brings us to the third problem: science has a fascination with the new (new data, theories, discoveries), and an ingrained skepticism concerning the old. Because of the history of science—steadily advancing toward unexpected horizons—science has often found that old assumptions (such as electricity being a fluid, the nineteenth-century concept of the ether, and so forth) turn out to have no basis in reality. The notion that an ancient, technologically backward society, an Asian society at that, could have discovered useful, perhaps universal truths about the mind using alien methods would be very difficult for many Western scientists to accept. However, it is important to understand that there is nothing in the scientific method that would preclude this possibility.

In reply to the second objection, it appears that the Yoga school of Hinduism has used "principles and procedures for the systematic pursuit of knowledge" (meditation), "involving the recognition and formulation of a problem" (as stated in the *Yoga Sutras*), and "the collection of data through observation and experiment, and the formulation and testing of hypotheses,"[8] (again, the process of meditation in conjunction with the master or guru, all with reference to a critical literature). As to science's emphasis on the "new," the third objection, science itself has often relied on the now-rather-old dogmas of scientific materialism to guide its search for the new. If, as Einstein and others have suggested, "it is the theory that decides what we can observe," then the "old," in terms of preconceived expectations, often plays an important role in the discovery of "new" knowledge in science itself.

Moreover, science has been in business for just four centuries. Perhaps it is only now beginning to mature to the point at which it can produce something as accurate as the *Yoga Sutras*. Who can say if Hinduism struggled awkwardly for five hundred years or more before making key discoveries and consolidating that new knowledge into a viable science of the mind, then spending a millennium or two testing and verifying its accuracy?

But if science's second and third objections (seeking expected results and relying on "old" knowledge) can be answered, what about the first one, the "lack of objectivity" in contemplation? How can yogic meditation be objective if it uses subjective methods?

REDEFINING OBJECTIVITY

As we saw previously, for science "objectivity" has historically meant distance or even, ideally, complete isolation from the object of scrutiny. Since humans' imperfect psyche was subject to all kinds of distorting influences, an objective process was needed to filter these out. Otherwise one fell into the pit of subjectivity, where the truth-seeker would be led astray by his or her imagination and biases. Therefore it was believed— naively as it turns out—that by using physical instruments and mathematics along with careful procedures of hypothesis, experimentation, and peer review, data purified of human subjective influences could be obtained. This belief was contradicted by quantum physics, which showed that the human mind is always involved in the outcomes of experiments. In addition, recent neuroscientific studies, such as those we have seen on the relationship between imagination and perception, underline the impossibility of achieving absolute objectivity in science or anywhere else.

In their explorations of the mind, contemplative traditions have used a different definition of objectivity. Since their "experiments" use a first-person, subjective approach, the notion of distance or isolation proposed in the scientific view of objectivity is irrelevant. Rather, many contemplatives believe that they can get a clear picture of mental phenomena by the skillful use of meditation techniques. The basis for that skill—using another, somewhat different, connotation of the word *objective*—is to be unbiased. One can be subjectively involved with phenomena—close, rather than isolated—without introducing distortions. Through intense and lengthy practice, the attention can be honed into a precision tool that, figuratively speaking, lights up the mind's interior. First one undergoes a sustained, rigorous training in developing stability and vividness of attention. One then uses one's enhanced powers of mental perception to learn to distinguish between the phenomena that are presented to the senses (including the sixth sense of mental perception) and the conceptual superimpositions that one under normal circumstances compulsively projects upon those phenomena.

Reaching that stage requires a long apprenticeship where detailed knowledge of how the mind operates, meditation techniques, and ethics play an important part—all within the overall aim of achieving spiritual goals. One example from such training in the Buddhist tradition describes four types of perceptual errors that may occur in meditation: (1) a defect in a sensory organ; (2) an objective basis for error, as in the cases of perceiving the illusion of a circle when observing a quickly turning firebrand; (3) a locational basis for error, as in the case of seeing trees moving from a moving ship; and (4) misperception due to distortions in one's own mental state. Such errors are pointed out in a detailed, systematic way in Buddhist literature.

For example, one of the most basic and important concepts of meditation—avoiding the extremes of laxity and agitation—is imparted in a well-known story of the Buddha (we paraphrase): A student who had formerly been a musician came to the Buddha one day for advice on meditation. The Buddha said, "I am told that when you were a musician you played the vina [an Indian lute] very well." The student answered, "Yes, that's true." The Buddha then asked, "When you tuned the strings, did you tune them loosely or tightly?" The student replied, "I tuned them neither too loosely nor too tightly, but in between." The Buddha replied, "It is the same with meditation. You should be neither too loose nor too tight."

At first glance this advice may not seem terribly profound. However, the avoidance of laxity and agitation is the principal gateway to sustained voluntary attention—the kind of attention William James, along with psychologists of the introspectionist school, didn't believe possible and that many Western psychologists and neuroscientists still don't. The Buddha was giving advice to students in 500 B.C.E., instructions at the novice level that reflected an understanding of the potentials of the

144 CONSCIOUSNESS—COMPLETING THE PICTURE

psyche that are still not generally recognized in the West. (As we shall see later, the use of functional magnetic resonance imaging on seasoned contemplatives appears to confirm their extraordinary abilities in sustained attention and other meditative techniques.)

Ethics, too, plays an important role in creating the unbiased attitude required for effective meditation. We saw references to this above from the Hindu yogic tradition. In Buddhism this attitude is typified by teachings on the "eight mundane concerns." The student is counseled to avoid concerning himself with gain and loss, pleasure and pain, praise and criticism, and fame and infamy. Rising above such preoccupations creates both psychological stamina and a disinterested attitude conducive to seeing things clearly. Here ethics helps contemplatives to avoid "cooking the data." Even the common ethical guidelines found in most spiritual traditions, such as not killing, stealing, committing sexual offenses, and so on, beyond their purely altruistic basis, help one maintain serenity. A guilty conscience is no more conducive to contemplative practice than nervous agitation or drowsiness.

So What?

Unless we choose to believe that contemplatives have been completely deluded and that they live in a dream world, the techniques and abilities shown above compel us to reappraise the meaning of empiricism. To do so, science must let down its guard a bit. We have already seen numerous examples of science's innate bias against the nonphysical. However, there are other strong prejudices that must also be overcome. Because science arrived to bring clarity to a world shrouded in mystery, knowledge associated with the mystical, vague, or mysterious has been deemed unreliable, if not pure quackery. Science has always considered quantitative features to be robust and viable, whereas qualitative aspects of life literally didn't measure up. How do you measure compassion or wisdom? How can you judge the capabilities of meditation masters if they are only to be found in distant mountain retreats? And what about all the strange terminology and poetic descriptions? They just don't "sound" scientific.

There are some very good reasons, however, why many contemplative traditions have kept themselves isolated from the hustle and bustle of society and have kept their most sacred and powerful practices secret. There have also been historical tendencies that have occasionally caused these same traditions to fall under the influence of dogmatism and to ignore their own principles of objectivity and empiricism. In the worst cases some contemplative schools have degenerated into hocus-pocus and quackery, giving mysticism as a whole a bad name, hence the common jokes about flying carpets and the rope tricks of swamis.

Unfortunately in the West there has been a tendency to apply such ridicule toward spirituality across the board. As we have seen, long-standing, clearly delineated contemplative traditions do exist and are becoming increasingly accessible to science and the West as communications shrink our world. It is no exaggeration to say that many Hindu, Sufi, and Buddhist traditions have made logical and systematic explorations of the mind resulting in experimental tools and techniques (of a first- and second-person nature), possess a critical literature, and sometimes incorporate academic debate. These are all hallmarks of an empirical tradition.

For its part, science has been susceptible to what historian Daniel J. Boorstin calls "illusions of knowledge." In his book *The Discoverers: A History of Man's Search to Know His World and Himself*, Boorstin characterizes the dogmatic belief that one already knows the truth as the principal obstacle to discovery.[9] We've seen how the beliefs of scientific materialism, held too literally and for too long, impeded the progress of modern physics and continue to impede neuroscience. As long as science accepts its own brand of empiricism, its own definition of objectivity, and its own tools of measurement as the only valid ones, it will miss a golden opportunity to impartially evaluate what may be a treasury of knowledge of the mind.

Empiricism is based upon experience. In the purest sense it demands confirmation of hypotheses by way of the senses. Advanced contemplative traditions absolutely require confirmation of their hypotheses, for without it the student could not advance along the spiritual path. For example, from the very beginning the Buddha counseled his would-be followers to "test my words as a goldsmith tests gold by heating, cutting, and rubbing." From another angle—one that is essential to contemplation—

empirical confirmation depends on the capacity of the perceiver. The existence of Neptune remained metaphysical or speculative until the proper instrument, the telescope, allowed for empirical confirmation. For astronomers, for whom Neptune had previously been a hidden phenomenon, the telescope widened their capacity to confirm the existence of objects in the heavens.

Contemplatives claim that the limitations on sustained attention are not absolute and that mental perception can be developed, in extraordinary and unexpected ways, into what is essentially a new instrument, a new capacity. This implies that there may be domains of reality that are metaphysical or hidden for science, but not so for contemplatives, to whom other types of experience and observation are available. Therefore, as soon as we add mental perception to our faculties for observing reality, the scope of empiricism must include the full range of subjective mental events, so often excluded or marginalized in scientific research. In this regard, contemplation, like science, can be said to be an empirical approach to understanding reality.

Where does this leave scientists and contemplatives? Both share an underlying belief that there is a difference between the way things seem to be and the way they actually exist. Neuroscientists have so far placed emphasis on the physical brain as the hidden source of mental phenomena. The motivation for such research is the accumulation of knowledge for its own sake as well as medicine's goal to help relieve suffering. On the other hand, contemplatives see the mind, which displays no physical characteristics, as central to human experience. "Outer" realities (the physical) thereby have a fundamental inner connection. The prime motivation for this kind of research, requiring subtle and detailed knowledge of the mind's activities and potential, is spiritual in nature.

If science can accept as a working hypothesis both the contemplative's definition of objectivity and the possibility and usefulness of sustained attention as a tool for exploring the mind from within, there is no reason why the two traditions cannot complement each other. The effectiveness of sustained attention as a tool to explore the mind could be tested by a combination of medical technology and third-person reports. (Indeed, as we shall see farther on, this evaluation process is now in its beginning stages.) There would also be benefits for contemplatives.

For example, Buddhism as it spread from India engaged each new culture at the highest level of discourse. As it comes to the West, it must engage this culture fully—which of course includes the scientific tradition—if it is to find acceptance. For its part, science could play a role in helping contemplatives to weed out untenable claims and ineffective practices. Anticipating this possibility, Tibetan Buddhism's Dalai Lama has stated firmly that if science can prove any Buddhist belief to be false, then that belief should be dropped. Such a criterion may be usefully applied to any contemplative tradition.

There is ample basis for believing that science and contemplation can together open and illuminate some of the black boxes of mind and matter. Contemplation can shed light on the nature of the nonphysical mind and its relationship to physical phenomena—a major stumbling block in both physics and the neurosciences. By the same token, the knowledge and tools of psychology and neuroscience can investigate the purported benefits of meditation, such as the refining of attention, the enhancement of wholesome emotions and attenuation of destructive emotions, the positive effects on physical health, and the creation of a greater sense of well-being.

Pioneering efforts have already been made. Ian Stevenson, Professor Emeritus of Psychiatry and former Director of the Division of Personality Studies at the University of Virginia, has conducted careful research over thirty years into accounts of children who allegedly recall specific people and events in their past lives—in other words, a scientific investigation of reincarnation.[10] Robert. G. Jahn, at the Princeton Engineering Anomalies Research (PEAR) Laboratory as well as other members of the program, has studied the mind's ability to influence physical objects. His research over a twenty-five-year period has demonstrated that consciousness and intention have definite and repeatable effects on physical objects, including various digital and analog processors (computers).[11] Until now research into such areas has been ignored by the larger scientific community, but the time has come for science to loosen the shackles of antiquated prejudice and to include both mind and matter within the world of nature, without reducing either one to the other.

Illuminating the Black Boxes of Mind and Matter 8

RECALL THAT the theories and calculations of classical physics, based on the work of Newton and his successors, are quite accurate for relatively large objects moving at relatively slow speeds. We can mount a satellite on a rocket and send it into orbit perfectly well using Newton's laws of motion and gravity. The rocket and its payload are large objects, and the maximum speeds obtained—in the neighborhood of 25,000 mph—are slow compared with the speed of light. Remember also, however, that the rules of classical physics break down when we enter the subatomic realm where objects are very small and velocities approach the speed of light. There, quantum physics and relativity provide far greater accuracy than classical physics (which in many cases can make neither heads nor tails of subatomic phenomena).

The classical and subatomic worlds can be viewed as two levels of reality, the former coarse and the latter refined. In science, physics especially, the latter is considered more definitive because physics seeks the ultimate, most detailed underlying causes of things. That is how the scientific search for truth has been defined for much of its history. However, as we've seen, this search has led to a paradoxical situation: at the subatomic level physical reality cannot be brought into sharp focus. Isolated causes and single, underlying mechanisms vanish into a broader, interdependent context where every item in a system plays a part, including the scientific mind that is searching. We are left with a scenario that reminds us more of a Zen koan or lines from Lao-Tzu's *Tao Te Ching* than anything written by Sir Isaac Newton:

Now what is the Tao?
It is something elusive and evasive.
Elusive and evasive!
And yet It contains within Itself a Form.
Elusive and evasive!
And yet It contains within Itself a Substance.
Shadowy and dim![1]

If quantum physics and relativity have led us to the edge of some kind of absolute truth—absolute in its very lack of absoluteness—there is more than a superficial resemblance here with a major theme in Buddhist philosophy: the "two truths."

LEVELS OF REALITY IN SCIENCE AND BUDDHISM

Buddhist philosophy was created with the practical aim of leading beings to the "awakened state"—the enlightenment achieved by Gautama Buddha and other realized beings. Therefore, even though Buddhists do take an interest in knowing and describing reality for its own sake (some Buddhist scholars even concentrate on this), the underlying motivation for this tradition is not essentially philosophical or scientific. What's echoed here is the ancient religious maxim "The truth shall set you free." Nevertheless, if these discoveries can shed light on science's search for truth, they are certainly useful and should be welcomed. The "two truths" of Buddhism may help a spiritual seeker on the road to enlightenment as well as a scientist, philosopher, or casual reader to a better understanding of how the universe ticks.[2]

The first of the two truths is *provisional,* conventional, or relative truth. It is relatively true that mind and matter, cause and effect, life and death, all of the things we commonly perceive and think about, exist. In our ordinary, conventional way of thinking, two things as different as the ground beneath our feet and our dreams for the future exist. As we saw earlier, such beliefs are accepted as definitive, absolute truth by some Buddhist schools, but for the Middle Way (Madhyamaka) school, considered by many to be the most profound Buddhist philosophical view, they are not absolutely existent. They exist only relative to a cognitive frame of reference—a mind-set—and not absolutely in their own right, independent of any conscious perspective. Just as high school and beginning undergraduate courses in physics concentrate on mastery of classical physics, Tibetan Buddhist monasteries begin with a study of the classical philosophies and beliefs in the true, inherent existence of phenomena, such as Vaibhashika, Sautrantika, and Chittamatra, before teaching the more definitive Middle Way.

While twentieth-century physics highlights the relative nature of matter and energy with respect to specific modes of inquiry, the Middle Way emphasizes the emptiness of all phenomena as independent things in themselves. Again be warned that *emptiness* does not at all mean "nothingness." Buddhist teachings on emptiness say that phenomena, commonly believed to have an essential nature in the sense that they are independent, inherent, permanent, and absolute—in other words, "really real"—are in fact "empty" of any such nature.

If I hold up my wristwatch and say, "This is *my* wristwatch!" there is nothing in the watch itself that makes it mine. The watch is empty of absolute mine-ness; there is nothing in its nature that makes it inherently mine. I could sell it, lose it, it could be stolen. Then I might no longer call it "mine," but the watch itself hasn't changed. Still, according to Buddhists, its appearing on my wrist or my saying that it is mine is accepted as *conventionally* true. Moreover, according to the Middle Way, the wristwatch itself is a mere convention. A certain collection of mechanical or electronic parts is given the label "wristwatch." Take it apart and there is no more wristwatch, even conventionally.

In order to function in the world at the conventional level, we accept the existence of wristwatches and our ownership of them. The first of the two truths still has value. However, there is a second, *definitive* view of truth. It does not deny that the provisional view exists and has relative value, in the same way that quantum physics doesn't deny that particles, fields, and waves as conceived in classical physics exist and interact with each other in statistically orderly ways. With classical physics alone one can, for instance, design a working automobile or the rocket mentioned earlier.

However, the relationship between the two levels of truth is critically important in the Middle Way. As we saw earlier, because the definitive level is *dependent* upon the first, or provisional level, the definitive level itself cannot be characterized as some absolute truth standing alone behind a flawed illusory view of things. When the watch on my wrist is carefully analyzed for any mine-ness and found empty of that, the *realization* of emptiness of mine-ness is itself dependent: if there were no wristwatch, analysis, or analyzing mind, no emptiness of mine-ness could be found there. Emptiness depends on something conventional being empty. This interdependence makes the two truths really one truth, two sides of the same coin.[3]

GOING TO EXTREMES

Hidden here is the potential for two enormous errors in thinking. First, if the definitive truth of "emptiness-of" is viewed as a real, operating-behind-the-scenes kind of absolute truth, then the conventional phenomena of provisional truths can be seen as "mere illusion," or as "nothing." A black-and-white, dualistic contrast is erroneously installed here between an absolute position and a nihilistic one. If you believe, as many scientists do, that the laws of nature are ultimate reality, then our everyday conventional, meat-and-potatoes reality could easily be viewed as pure illusion or as something secondary or less real than the ultimate reality that supports it or from which it emanates. When we examined the wristwatch, we found it empty of any quality that made it "mine." But in observing it, we also did not discover some "universal law of emptiness" supporting the watch's existence. A wristwatch is simply empty of any inherent identity of being a wristwatch (we have merely given its parts a label), let alone being "my" wristwatch. Further, that emptiness is not some deeper reality underlying the wristwatch itself. Emptiness is not the sort of ultimate reality sought by metaphysical realism within the philosophy of scientific materialism.

Second, if we slip all the way over to the other extreme and view emptiness as nothingness or pure illusion (rather than "emptiness-of" some quality), then phenomena—mind and matter—are given a contrasting concreteness. It becomes a case of "all versus nothing": the phenomenal world becomes the only reality. In this case we have given the first truth too much emphasis. We have made commonly perceived phenomena definitive, absolute, independent, inherent, and permanent—"really real." That would be like wholly believing that there is something about the wristwatch itself that makes it a timepiece. According to the Middle Way, in both cases we have gone too far. Something in our habitual way of thinking and perceiving has made us extremists. We give phenomena either too much reality or not enough.

Doesn't this sound familiar? Here we have two positions: The first, which places emphasis on underlying causes, is commonly found in Western philosophy and science. The second, which considers phenomena to be really real, is our ordinary, "surface level" view of reality. If the emptiness of inherent existence in phenomena is made into an absolute truth—the first mistake—this is similar to the assertion by metaphysical realists that there are absolutely objective mechanisms underlying the observable universe. The Buddhist theme of emptiness—which appears to have some resemblance to quantum physics regarding the subatomic world—then becomes a transcendental "scientific law" operating behind the scenes. We have fallen into the error of imagining another universal clockwork for explaining reality, this time based on a reified notion of emptiness. The second mistake—our normal, naive view that the only valid reality is the apparent, observable universe of our senses—argues that any analyses undermining it are meaningless, "mere illusion." We bang our fist on the table, point to it, and say, "*That's* real!" Any analysis of the subatomic basis of the table (or the Middle Way's emptiness of permanent, inherent existence of the table), we call "pure baloney."

Today's science accommodates both the utility of classical physics and the greater overall accuracy of quantum physics and relativity. Classical physics functions in the macroworld even though it is still subject to subatomic laws drawn from quantum physics and is most accurately described by them. Similarly, the Middle Way accepts conventional reality for what it is—appearing phenomena—along with its definitive character, emptiness. Both modern physics and the Middle Way imply that conventional realities of all kinds, including space, time, and matter, are all empty of true, inherent existence. Yet if they assert that matter is empty of inherent existence, that it isn't really real, exactly how could such a universe, unreal in commonsense terms, function at all? If it isn't real, how come it works?

EMPTY—BUT NOT ZERO

Because we are in the habit of thinking that physical entities are absolutely real, we are automatically skeptical of the notion that any object that is less than that could function in our world, a world of "hard facts." We have been trained since childhood to believe in simple cause and effect. We believe that a cause has some power to affect something else. That is how we define *cause*. These in turn create real sequences that comprise real events. If A affects B, then A must have some real, inherent quality that is its power to do so. Gas from the tank is injected into the cylinders, where explosions drive the crankshaft connected to the gears connected to the drive train connected to the wheels and voilà: the working automobile. The reality of such causal sequences along with the absolute reality of all of the parts in the sequence has become part of the seeing we believe in. However, we know that things aren't always the way they appear. They appear the way they do to us due to a lot of mental baggage and fancy packaging. So maybe we're now in a position to give emptiness a chance.

Contemplative traditions that advocate various views of emptiness, such as Buddhism and Taoism, use a variety of analogies to break down our realist tendencies. One example is the story of the seed and the sprout. If we plant a seed in the ground, at some later time a sprout will often appear. What's going on here? Is the seed the cause of the sprout? We may be tempted to say yes. That would follow everyday cause-and-effect reasoning. We saw only a seed when we did the planting. A week or so later, when we look at the same spot we see a sprout but no seed. Yet how can one thing produce another?

In the following pages, we will present a methodical analysis that is often used in Buddhist philosophical texts to deconstruct our notions of the inherent nature of causality. This approach works with the following four possibilities

1. The sprout could be self-caused, that is to say the potential sprout is in the seed to begin with.
2. The sprout could be inherently caused by something other than itself. The seed, as something absolutely distinct from the sprout, is the cause of the sprout.

3. The sprout could be both inherently self-caused and other-caused. A combination of both its potential in the seed and other causes such as moisture, sunlight, soil, and proper temperature intrinsically cause the sprout.

4. The sprout has no cause; it just appears with no prior causes.

UNDER THE SPELL OF COMMON SENSE

Unless we believe that everything occurs randomly, we can dispense with the fourth possibility. When contemplating the other three options, the Middle Way carefully examines the meaning of the word *cause*, so let us look first at what a cause might be. If we take that to mean something absolutely real, this implies that causes are distinct and independent— some piece of the universe we believe to be separate from the whole. Usually we do understand a cause to be some real "thing" that we can identify as it and it alone. To believe in that kind of reality means to believe that the universe is made up of distinct, really real, things. That's what we normally do. We call it common sense. However, as we have seen, this is based on a materialism that many scientists no longer support.

How can we reason about causes? Suppose I depress the letter *x* on my computer's keyboard. A letter *x* then appears on the computer screen. We would normally and naturally call my depressing the key the cause of the letter's appearance on the screen. But suppose instead of punching the key I scratched my nose (and no *x* appeared)? Would that be a cause of the *nonappearance* of the letter *x*? Or if I scratched my nose with one hand while punching the key with the other, would scratching my nose still be a "noncause" of the appearance of the letter *x*? If there are such things as causes, mustn't there also be noncauses? Aren't all of these options based on the same reasoning?

Or suppose we said, "The king had the power to execute any of his subjects. But he was powerless to bring a dead subject back to life." When viewed that way, powers and causes (and powerlessness and noncauses) seem to take on a life of their own. One can have the power to do something or be powerless to do something, but where is a real "power" or "powerlessness"? No one has ever seen a power or a cause leaping like a ghost from finger to computer screen, from king to executioner, or from the gas tank to an automobile's wheels. Yet we believe in and imagine them so strongly that they are conjured into existence by our merely thinking of them.

The Middle Way proposes an alternative explanation for the appearance of the phenomena of the universe—regularities. Certain things tend to occur together or in sequence. Whereas causes imply to us some power to affect, the Middle Way defines appearances as mere regularities. In this view the "chain of events" leading to an automobile's movement, or a seed's "becoming" a sprout and then a plant are only regularities. The gasoline, spark plugs, cylinders, crankshaft, and so on in a car are regular events in the movement of an automobile. If it sounds unlikely, even crazy, that the movement of an automobile has no inherently real cause, is this because we have proven knowledge that the universe and its causes must be absolutely real? Or could it be because we are under the spell of a limited view that we have never deeply investigated?

DOWN ON THE FARM

Now let's investigate the second possibility listed above: the sprout is inherently produced from something other than itself. A seed is not a sprout. However, we can trace intermediary stages between a seed and a sprout by making observations at short intervals. On day 2 the seed has opened. The next day there are thin tendrils emerging from it, then roots, and so on. There is a continuum here between seed and sprout that we artificially fragment with our conceptual labels of "seed," "sprout," "plant," and so on. If we extend that continuum backward and forward in time, we recognize that our "seed" was once part of the seedpod of a mature plant and our "sprout" is just a stage before becoming a plant.

We have taken a fluid process and chopped it up according to the practical needs of farmers and gardeners: a seed is a stage in a process that the farmer can easily store, tote around, and put into the ground. A sprout is a sign that his plant is growing, that the seed was healthy, and that the necessary conditions for growth are being met. Yet there is nothing in that little kernel that is a static, permanent, absolute "seed." Nor is the sprout or the mature plant anything permanent or absolute. All are part of an interdependent continuum or process, empty of an absolute identity.[4] We conceptually superimpose on experience our cookie-cutter concepts, thereby creating "things" that "have" their own parts and qualities. However, apart from our concepts, which create this dichotomy, no phenomenon "has" another phenomenon as its characteristic. Furthermore, apart from the attributes we mentally apply to the various things we identify, no essence nor any real "thing" that has those qualities is ever found. My wristwatch, alluded to earlier, not only lacks any inherent quality of being "mine," it lacks, or "is empty of," any such absolute identity in and of itself as a wristwatch. That absence of an essence is what is meant by "emptiness."

LET'S HAVE A RELATIONSHIP

Is that really so far-fetched? We can easily admit that words, descriptions, and qualities such as "seed" and "sprout" are merely labels—useful conventions rather than absolutes. Yet what about the stage "seed" being the *cause* of the stage "sprout"? Or are all of the microstages between seed and sprout taken together the cause of the sprout? No, that wouldn't work, because what, then, would be the cause operating between one microstage and the next? Even if we carefully examine the seed with a powerful microscope at intervals as short as a minute or two, we know that there are always changes going on at the cellular and molecular levels. Therefore, what is causing these changes?

No matter how we chop up the stages between seed and sprout, no matter how closely we examine these with our most powerful and sensitive instruments, we discover no "power" that causes one thing to turn into another.

If we conceive of one stage as an absolute, permanent, independent entity, by definition it cannot have any relationship to anything else. By definition, two completely self-contained, independent, permanent, absolute things cannot affect one other. If they did, they wouldn't be self-contained, independent, and so on. But if we back off that position and say that there is simply a "relationship" between them, Middle Way philosophers will point out that we are now viewing these things (such as seed and sprout) as relative, conventional realities. A relationship composed of regularities doesn't require absolute realities or absolute causality, and the relationship itself lacks any such inherent existence independent of the things that are related. Seed and sprout and their causal relationship, though existing conventionally, are now seen as being "empty of" absolute existence.

How could there be, even in principle, a "power" to cause an effect? Such a power would also have to be a thing, absolute and independent. Yet where could such a power originate? The cause of that power would have to be another power, which would have to have a causal power behind it as well. The whole chain, going back infinitely, would be conventional and relational, not absolute. On the other hand, if it had no origin, how could it exist at all? Intrinsic powers that function as causes are never to be found, they are simply assumed to exist according to common sense. They are conventional labels useful at the relative or provisional level, no more. Causes and origins exist, but only at the conventional level.

No Plant Is an Island

Now, what about our first option, self-causation? Is there some potential that we might label "sprout-ness" in the seed? The notion of the sprout being potential in the seed is similar to themes that crop up in genetics. A biologist would say that the seed contains the genes of the sprout (and of the whole plant). Since genetic material is found in all phases of the plant, this appears at first glance to be an example of self-causation. However, do the genes have some inherent power of their own, independently of other factors, to cause the sprout? Genetics tells us that genes are "chemical keys," where chromosomes made of DNA and proteins play a part in unleashing a pattern of development over time. Yet when genes are observed and analyzed in the laboratory, are any powers actually detected? At the molecular level, complex chemicals are shown to combine, evolve, and so forth, but this is just one more continuum. What causes one chemical to combine with another to form a more complex chemical? Where is the power, the agent that causes this to happen? Again we come up empty. We find only regularities. In addition, if we were to continue our examination on down to the subatomic level in our search for an agent, we would run up against the limits imposed by quantum physics: our subatomic agents fade into the quantum world of quasi-existence.

When we consider causation by another entity, we do detect conditions that occur regularly in conjunction with an event. The growth of a plant usually involves a seed, sprout, and mature plant, in that order. The movement of an automobile involves the set of regular conditions described above. However, analysis of the conditions of any cause-and-effect sequence never yields any causal power in either the conditions or their interactions—and in science, just like in the Middle Way, analysis is the name of the game.

The last remaining option, causation by both self and other, can be exemplified by describing the sprout creating itself (self-causation) aided by contributing causes such as water, sunlight, and so forth (causation by other). Yet since no real, causal power was found in either self-causation or causation-by-other, there is no possibility that they might appear in combination. An absolute power to affect can no more be found in a combination than it can in a causal sequence.

Where does this leave us? What kind of existence is this where things function but nothing is inherently real? If the Middle Way is correct, this leaves us in the world *as it is*, the very world humanity—and science—has been trying to fathom for millennia. Common sense keeps telling us that the "things" we constantly and compulsively separate from the flow of life are absolutely real: independent, inherent, essential, and permanent—with some essence that makes them "what they are." They seem that way because we constantly put energy into labeling them that way. However, common sense goes way overboard in attributing such an independent reality to things.

The question is not, How can such a world characterized by the emptiness of absolute reality function? Rather, the question is, How could things function if they were not empty? Imagine an automobile in which all the parts were absolutely real. The gasoline could not be turned into the hot gas of the explosion that drives the cylinder. The spark plug could not suddenly emit a spark to cause the explosion. No electricity could even flow to the spark plug. Nothing could move or interact! However, because these parts are relatively and not absolutely true, they can morph into the various forms necessary for the wheels of existence to turn.

It is precisely because the conditions that surround events are empty of absolute reality that events can take place at all. This gives them the flexibility to be interdependent. According to this view, interdependence—the interrelationship among things that characterizes existence—accounts for the very structure, coherence, and intelligibility of the universe. The emptiness of absolute phenomena, of underlying mechanisms, real causes, immutable laws of nature, and so on, merely shows that such an inherent identity for things cannot be found under analysis. In other words, they don't exist aside from the conceptual mind-sets in which they are conceived. Regularity is all there is to it! The Middle Way provisionally accepts the existence of the conventional world, but as empty of absolute reality. Realism, however, seeks some absolutely real, underlying foundation for conventional reality. According to the Middle Way, it is from this dream that we must awaken.

TWO BLACK BOXES—OR ONLY ONE?

If contemplative traditions such as Buddhism shed any light at all on the relationship between mind and matter, the message is that the two are relational, interdependent. The black box of matter, of so-called physical phenomena, does not contain a bunch of absolutely real, permanent, independent objects, no matter how small. We cannot know what they "really are" because in that sense there is no "really." Yet they are not "nothing," because they do appear and function. They are present in a manner that is neither absolutely existent nor absolutely nonexistent. According to the Middle Way, if we stop imposing our realist and nihilist views on phenomena, we may be able to see directly what they are. That is considered a spiritual accomplishment, and its realization is said to be beyond the power of words to express. (Stories from Zen Buddhism sometimes describe one's reaction to such realizations as a hearty laugh.)

There is clearly a parallel here between major themes in subatomic physics and the Middle Way's view of physical phenomena. Just as subatomic particles cannot be found to be inherently real—existing in isolation from the scientific mind that measures them and the instruments designed by scientists—sprouts, seeds, automobiles, games of professional basketball, and everything else are, to the Middle Way, merely labels for interdependent processes in a participatory universe. Just as an electron is empty of the simultaneous position and momentum required for its existence as something real in classical physics, under analysis no independently real power (in commonsense terms) can be found that transforms a seed into a sprout.

Now, is there any similar parallel between neuroscience and the Middle Way? What's in the black box of the mind? As we've seen, the major thrust of neuroscience is to seek physical causes for mental phenomena. Some neuroscientists (and philosophers of mind) see the mind as an emergent property of the physical brain, others say the mind *is* the brain or simply a function of the brain and that therefore there is no need for the term *mind* at all. The choice is between the subjective mind being an illusion and its being a nonentity.

From a Middle Way perspective, materialists err on both of the extremes to be avoided: they state that the brain is substantially, inherently real and that the mind is an ineffectual illusion. This devaluation of the mind is their solution to Descartes's error of viewing both the brain and the mind as independently existent substances, mysteriously joined by the pineal gland. However, as we have just seen, according to the Middle Way, if they did exist independently, there could be no causal interactions between mental and physical phenomena. Neuroscientists have focused their attention on the brain and see it as real—they project reality upon it, label it as "real." Therefore, since they can't measure mental phenomena directly, it is quite natural for them to view the mind as relatively unreal. However, this bias is an inevitable outcome of their lopsided ways of inquiring into the mind-body problem.

Earlier we took note of the problem of attributing to the brain the causal power of creating the mind—consciousness, thoughts, images, feelings, and so on. This is the so-called hard problem encountered when we try to find a link between mental phenomena and their brain correlates. Thoughts, emotions, feelings, perceptions, and so on are imagined to be caused by (or actually exist as) the firing of neurons and changes in brain chemistry. According to the Middle Way, the "hard problem" is really an insoluble problem as long as you view either the brain or the mind as inherently real. Examining this problem even more carefully, the Middle Way might say, "If an electrochemical process is an absolutely real activity, how can it 'become' an illusory mental phenomenon with the nature of consciousness? How can anything considered absolute 'become' something else?" Or phrased another way, how can a collection of electrochemical processes become the very consciousness that supports or is aware of all the varied kinds of mental phenomena we experience? How can neurons and neurotransmitters "make" me aware or "become" awareness? Where is that causal power?

If electrochemical processes are absolutely real, they must be isolated and therefore cannot interact. If they cannot interact, they cannot affect some other absolutely real, independent entity, such as a neurotransmitter being the absolute cause of some physical change in brain tissue. We have the same problem describing the brain's physical activity as we would with an automobile made up of absolutely real, independent parts and processes. Again, we can examine the brain's electrochemical processes from the cellular on down to the molecular and then the subatomic level and we will never find a bridge of cause and effect between supposedly absolute, independent entities. We merely imagine such inherently real connections with our deeply ingrained habits of thought, strongly influenced by metaphysical realism. According to the Middle Way, they have no such absolute existence. However, that does not prevent neurons and neurotransmitters from existing as regularities that form part of an interdependent system that in some way influences the generation of states of consciousness.

Furthermore, if the brain is considered an absolute, independently real entity, haven't we simply placed a label on a bunch of parts: neurons, synapses, blood vessels, and other kinds of specialized fluids and tissues? Are not the nerve cells in the tips of my fingers part of the brain? Is the wetness of water I sense through the nerves of my skin absolutely separate from my tactile receptors in the brain and my consciousness of that sensation? Is the appearance of an object, say a tree, seen in my visual field truly independent from my visual cortex and my consciousness? If no absolute, truly independent identity can be attributed to any phenomena, how can an absolute difference between entities be asserted?

Therefore, according to the Middle Way, the physical brain is not an absolute. It exists as something conceptually designated upon its parts and functions. On the other hand, the mind is not a pure illusion or a nothing. Although that which is designated as "the mind" is conventional and therefore dependent upon its attributes (thoughts, images, feelings, and so on), this *concept*, "the mind," does appear to our consciousness. Such a phenomenon, often nebulous and hard to grasp, does exist. If we didn't perceive and experience the mind, how could we conceive of it or talk about it? Our mental experience is just as real as a physical brain preserved in a jar of formaldehyde. Furthermore, the mind interacts fully with the so-called physical world. Our mind moves our hand to lift a cup of tea. Using his mind, Henry Ford thought up the assembly line, and before long the world was flooded with working Model T Fords.

A RAINBOW STATE OF EXISTENCE

Are the black boxes of mind and matter really separate? Truly existent, absolute, independent matter is not to be found. Nor can we find such a mind. Is it safe even to say that phenomena are experienced by consciousness? Or does our knowing, stripped of all of its distortions, allow us to say only that phenomena are experienced, with no independent "experiencer" implied? If that is the case, then everything boils down to the intimate act of contact we call "knowing" or "consciousness."

However, let's descend from such speculative heights and once more use the rainbow to put things into perspective. The rainbow can be called a physical phenomenon in that it requires moisture (H_2O, to the world of chemists), sunlight (whether waves or photons, still an enigma to physicists), and physical organs—the human eye plus brain. It can be called a mental phenomenon because it doesn't exist apart from the mind that perceives it. A rainbow is not absolutely real, because something other than the interdependent connection of moisture plus sunlight plus the proper viewing angle for a mind cannot be found. Where among these is the "rainbow itself"? However, it is not nonexistent either, because it can be seen and it functions in the world. For example, it causes people to smile and say, "How beautiful!" Photographs of rainbows are taken and developed. (The state of Hawaii is called the "Rainbow State" and pictures of rainbows are printed on its license plates and travel brochures.) According to the Middle Way, mind and matter exist in the same manner as rainbows: they appear to consciousness (and therefore exist), but are empty of absolute existence. Everything exists in that way. So the black box of mind and matter contains, figuratively speaking, only rainbows.

Science and Contemplation 9
Probing the Mind Together

W E NEED a better understanding of the mind. That goes for those of us who follow a spiritual tradition as well as those with no such interest at all. For the former, the mind is fundamental for a clear understanding of spiritual beliefs and experiences. Whether the source is Jesus, Muhammad, Brahma, or Buddha—whatever the revelation that inspires us—a mind made clear by self-knowledge will be better able to understand the message than one immersed in confusion. Similarly, from the secular side, it is obvious that when we ignore the mind from which all of our views and decisions arise, we get into difficulties. These range from our own emotional and interpersonal problems on out to the strife and suffering of war and terrorism. Our world is essentially mind-made, and we are therefore in dire need of knowing our minds better if we are to prosper or even survive as a species.

Science, for reasons we've seen previously, is a relative newcomer to the study of the mind. It has made significant progress in understanding the physical correlates of mental phenomena (the brain) and it has made many sound inferences about the mind. Yet no matter how much technical data neuroscience can extract from the brain, it has not been able to fathom the central issue, consciousness, because it does not know the mind intimately from the inside, first-person viewpoint. In contrast, the world's contemplative traditions have little to say about the brain but have investigated the mind over several millennia. Knowledge of the mind is the forte of Buddhist contemplation.[1] Therefore, when it comes to the exploration of the mind-brain nexus, science and Buddhism naturally complement each other.

As we shall see below, Buddhism and science have begun to probe each other, checking to see whether a useful relationship is possible. Buddhism, as it assimilates into Western culture, is being forced to reexamine its tenets and meditative technologies from a Western standpoint, which must include scientific principles. If science can confirm certain aspects of Buddhist psychology and cosmology, Western Buddhists, in particular, will gain greater confidence, and people openly curious about contemplative traditions will find encouragement. From another angle, all spiritual traditions, Buddhism included, acquire over time a residue of dogmatic beliefs, superstitions, and purely cultural elements having little to do with the advancement of spiritual knowledge among their practitioners. Buddhists' involvement with science may act as a cleansing influence to remove some of this tarnish.

As an empirical method of investigation, science shares the same approach the Buddha expressed about his teaching: theories must be empirically and rationally investigated, not merely accepted and turned into dogma. In the words of the Dalai Lama:

> When investigating the ultimate nature of reality, Buddhist thinkers take the Buddha's words not so much as an ultimate authority, but rather as a key to assist their own insight; for the ultimate authority must always rest with the individual's reason and critical analysis.[2]

Thus, there seems to be a reasonable chance that science and Buddhism together can advance our knowledge of the mind-brain while providing answers to questions vital to each tradition.

WHITE COATS AND RED ROBES

Actually, this collaborative process between scientists and contemplatives is well under way. It began in the late 1960s when doctors from the Harvard Medical School and the University of California at Irvine measured the effects of meditation. They found that a group practicing Transcendental Meditation showed a 16 to 17 percent decrease in metabolism (oxygen consumption). This physical change resulted from the subjects simply changing their thoughts. Other effects included a decrease in the breathing rate and a lowering of blood lactate, indicating diminished stress and anxiety. The researchers theorized that two activities—repetition, as in the recitation of a word or prayer, and deliberately disregarding competing thoughts—led to a "relaxation response." As a result of similar studies over several decades, meditation of this sort is now a recommended treatment for hypertension, cardiac arrhythmias, chronic pain, insomnia, and the side effects of cancer and AIDS therapy.

One of the pioneers in the application of meditation to stress is Jon Kabat-Zinn. In the late 1970s Kabat-Zinn, who has a PhD in molecular biology and is a longtime practitioner of meditation, founded the Stress Reduction Clinic at the University of Massachusetts Medical Center. Due to the clinic's dramatic success in helping patients heal from a wide variety of illnesses, his Mindfulness-Based Stress Reduction program is presently taught in over 250 clinics throughout the world. Kabat-Zinn continues to do research on the relationship between meditation and health, and his book *Full Catastrophe Living* has become a bible of stress reduction and meditation-based healing.

In the 1970s and 1980s teams led by Herbert Benson, a professor of medicine at Harvard, traveled to northern India and Sikkim to study Tibetan yogis practicing a form of meditation called *tummo*. In this technique body heat is increased dramatically through a specific meditative practice. This has both a spiritual and a practical advantage: the mind is purified while the increased heat allows yogis to be comfortable over long periods of time in frigid caves and huts at high altitudes. The scientists witnessed naked yogis in freezing temperatures drying wet towels draped over them by using their increased body heat alone. One practitioner showed an astounding 64 percent drop in oxygen consumption. Others showed increased oxygen consumption. In either case, body heat, which is normally directed away from the skin when we become cold, was manifesting on the body's surface. Nothing of its kind had been observed previously by medical science.

THE LATEST FROM THE LAB

Studies such as these show that techniques of meditation do cause dramatic physiological changes in the brain and elsewhere and that they can be measured by scientific instruments. Dr. Richard Davidson of the University of Wisconsin at Madison began studying the relationship between emotions, the brain, and meditation in the 1970s. He now heads the W. M. Keck Laboratory for Functional Brain Imaging, a state-of-the-art laboratory where he has recently begun studies of the physiological effects of emotional states generated by meditators, including some who are very experienced. His main tools are fMRI (functional magnetic resonance imaging) and an advanced electroencephalograph (EEG), a web of up to 256 electrodes that allows for great accuracy in detecting the source of electrical messages within the brain. These tools permit him to observe and videotape changes taking place in the brain in real time. For example, a subject can be asked to generate a certain state or initiate a specific meditation practice and report on his or her experience while the researchers closely track the brain's physical changes.[3]

A radically different approach to exploring mental functioning has been employed by Dr. Paul Ekman, a professor emeritus of psychology at the University of California at San Francisco. Ekman has developed the Facial Action Coding System, a refined method for determining the abilities of human subjects to detect subtle changes in facial expressions and the emotions behind them. FACS is now used by law enforcement agencies, including the Secret Service. Several years ago Ekman became curious as to whether highly trained meditators might exhibit greater sensitivity to very brief and subtle facial expressions than the normal population. He also conducted experiments on meditators where he assessed their startle reactions to sudden loud noises. What have Davidson, Ekman, and others discovered so far?

SURPRISING RESULTS

One of the key subjects in these preliminary investigations has been a Western monk highly trained in a variety of Tibetan Buddhist meditation techniques, but with a strong background in science. In Ekman's research this monk (code-named Osel) and several other seasoned meditators produced extraordinarily high scores in their ability to accurately detect the emotions behind an array of facial expressions. Ekman believes this indicates a high degree of compassion and emotional sensitivity.

Furthermore, Osel was able to buffer the startle response to loud noises to a degree Ekman had never seen before. This demonstrates an ability to powerfully concentrate the attention.

Osel was also measured while activating on command six distinct meditative states (including one-pointed concentration, visualization, pure awareness, and compassion). The differences among these states were clearly observed on Davidson's instruments. For example, when Osel generated compassion, a dramatic increase in electrical activity was seen in the left hemisphere of his brain—a region known to light up when subjects are experiencing positive emotions. Again, the results were far beyond any Davidson had measured previously. Similar data have been observed by Davidson's research on several other advanced meditators.

Of course this particular subject may be unusual in himself: his high scores might indicate genetically inherited talents having nothing to do with his training in Buddhist meditation techniques. It may also be the case that those attracted to long-term meditation are "naturals" like him. New research is now under way to expand these studies to larger populations of both experienced and beginning meditators. One is being conducted in a collaboration between the University of California at Davis and the Santa Barbara Institute for Consciousness Studies.

Davidson and Kabat-Zinn have already teamed up on research into the effects of meditation on a slightly larger population. Davidson characterized as "tantalizing" the results from this study of highly stressed workers for a Wisconsin biotech firm. Company employees to be examined were selected at random as to whether or not they would receive eight weeks of basic training in meditation, thus providing a control group consisting of the nonmeditators. Preliminary results demonstrate that those who received this training showed an increase in left prefrontal lobe activation, both at rest and when presented with emotional challenges. Again, increased activation of this area of the brain is associated with positive emotions and reduced stress, along with improvements of the immune system. Preliminary results from a similar study of classroom schoolteachers conducted by the University of California at San Francisco and the Santa Barbara Institute show a lowering of levels of depression.

The Mind and Life Institute

A key organization promoting this kind of research is the Mind and Life Institute, which also cosponsored the study at UC San Francisco. MLI was the brainchild of Francisco Varela, a Chilean neuroscientist and phenomenologist based in Paris, and R. Adam Engle, an American businessman. Both had backgrounds in Buddhist meditation and approached the Dalai Lama about organizing an exchange between Buddhists and scientists. For his part the Dalai Lama has maintained a lifelong interest in science, to the extent that he once said that if he hadn't become a monk he might have chosen to be an engineer.

The Mind and Life Institute, which held its first conference in 1987, has developed a format whereby a topic is chosen, and leading scientists in the field are then invited to make presentations and discuss them with the Dalai Lama over a period of five days. Topics have included modern cognitive science; consciousness; emotions and health; sleeping, dreaming, and dying; altruism, ethics, and compassion; physics and cosmology; epistemological questions in quantum physics and Eastern contemplative sciences; destructive emotions; and neuroplasticity. Among the participants have been such eminent scientists as the physicists Anton Zeilinger, Arthur Zajonc, Steven Chu, and David Finkelstein; the astrophysicist Piet Hut; the biologist Elizabeth Blackburn, recipient of the Lasker Prize; the geneticists Eric Lander and Fred H. Gage; the neuroscientists Antonio Damasio, Robert Livingston, and Helen Neville; the medical researcher Dean Ornish; the psychologists Ervin Staub, K. Anders Ericsson, and Nancy Eisenberg; the computer scientist Newcomb Greenleaf; the economist Robert Frank; and the philosopher Michel Bitbol, research director of the Centre de Recherche en Epistémologie Appliquée in Paris. Distinguished spokespersons from spiritual traditions, the humanities, and social sciences have also participated.

A Renaissance in the Offing?

There are numerous potential benefits for such cooperative research between scientists and contemplatives. To begin with, a deeper, more accurate understanding of the relationship between the mind and the brain will have a dramatic impact on key issues in science, spirituality, and philosophy. The existence or not of a mind (or soul) with some degree of independence from the physical brain has direct implications for the question of free will and moral responsibility, not to mention such topics as reincarnation and psychic powers. Free will is a central question in both philosophy and religion, and reflects on everything, including the ethics of our personal relationships and the organization of society as a whole. It is closely connected with the notion of personal responsibility: to what degree and in what manner are we responsible for our actions? The existence or not of free will has been argued throughout human history, but now, with combined research examining its inner/mental and outer/physical aspects, there is a possibility of clarifying this vital issue. Obviously this will take a great deal of careful thinking and research.

The scientific examination of meditators has helped change neuroscientists' views on neuroplasticity, the ability of the brain to restructure itself from experience. Formerly it was believed that the neural patterns and connections in the brain were relatively fixed in the adult human, but degraded significantly due to aging. We now know that in fact the brain can grow and redesign itself through training and other kinds of conditioning, even later in life. Mental approaches to sports training and conditioning as well as meditation have proven this adaptability. Mind (in the sense of subjectively experienced mental processes) definitely affects the gray matter of the brain. Concentration and visualization techniques can improve your golf game or make you a calmer person. (Professional basketball coach Phil Jackson was innovative in the use of such techniques during his tenure with the Chicago Bulls, who dominated the National Basketball Association in the 1990s and were known for their spectacular skill and cohesion.)

Despite the mental component of neuroplasticity, there remains a danger in emphasizing its physical side. Neuroscience, as we have seen, is still heavily biased toward the physical, and the very word *neuroplasticity* could merely replace the old hardwired view of the brain with a more flexible physical image—self-transforming electrochemical goo. From there the temptation to favor drug treatment, electroshock (electroconvulsive therapy), or surgery over therapies that can change our mental environment reasserts itself.

There are enormous potential benefits in cooperative research between scientists and contemplatives for physical and mental health. Is it possible that certain diseases could be treated effectively with forms of relaxation or meditation? Such tools are common in traditional Tibetan, Indian, and Chinese medicine. There, as touched on previously, a system of "channels" associated with the physical body is sometimes accessed through techniques such as visualization to effect a cure. These channels have not been observed by Western anatomy and are not considered physically existent by the doctors who utilize them. Even so, acupuncture, which has proven its effectiveness for a variety of illnesses, is based upon these invisible patterns.

Tibetan and other Asian medical traditions also employ complex herbal formulas to treat disease. One of the principles involved in producing these medicines is the neutralization of unwanted side effects. Western medicines often attack diseases quickly and dramatically, but they also can have strong side effects, especially when the patient is taking several drugs that may interact unfavorably. This is one of the great disadvantages of Western pharmaceuticals. New problems with specific medicines and their interactions are chronicled often in the press, with the U.S. Food and Drug Administration frequently issuing new directives for drugs previously thought safe. In contrast, Tibetan medicines often contain fifty to a hundred separate ingredients, some of which are intended to neutralize any side effects. How did Tibetan doctors come up with such complex formulas without the aid of chemical laboratories, test animals, and the other accoutrements of modern science? In addition to the empirical testing of pulses and urinanalysis, the source of these formulas is also mental: many of the formulas have been allegedly received in meditation and dreams from divine emanations such as the Medicine Buddha.

CONTEMPLATIVE PSYCHOLOGY

How well do these remedies work? The Tibetans, Indians, and Chinese who use them appear to have confidence in them. In China traditional medicine is offered alongside Western medicine, sometimes in a complementary fashion. Hospitals provide both kinds of care, and doctors of traditional medicine study in the same institutions as those learning Western medicine. Indeed, the Chinese population generally prefers traditional medicine over the Western brand. A respectful dialogue between Western and Oriental health care professionals would provide the opportunity to test both the medicines and the philosophies behind them.[4] The benefits could spread both East and West, creating a renaissance in the healing arts.

The most obvious area of potential benefit in a dialogue between science and spirituality is psychology. It is not surprising that contemplative traditions have such a strong interest in this field. These inward-focused systems view the mind as the pathway and (when it is dysfunctional) the major obstacle to spiritual accomplishments. If an agitated, dull, or unbalanced mind blocks spiritual understanding and contact with higher realities, especially if such negative conditions are viewed as typical of our normal, unenlightened state, then improvement based on a detailed understanding of the mind is essential for spiritual progress. Therefore, as we have seen, the written and oral teachings of contemplative traditions, such as Hinduism, Taoism, and Buddhism, are filled with detailed information on the mind gleaned from many centuries of study and meditation. A mere hint of the subtlety and depth of this knowledge is given in a paragraph written by the Dalai Lama in a recent book:

> There are explicitly cognitive states, like belief, memory, recognition, and attention on the one hand, and specifically affective states, like the emotions, on the other. In addition, there seems to be a category of mental states that function primarily as causal factors in that they motivate us into action. These include volition, will, desire, fear, and anger. Even within the cognitive states, we can draw distinctions between sensory perceptions, such as visual perception, which has a certain immediacy in relation to the objects being perceived, and conceptual thought processes, such as imagination or the subsequent recollection of a chosen object. These latter processes do not

require the immediate presence of the perceived object, nor do they depend upon the active role of the senses.[5]

This is but a general picture of the mind from the Tibetan Buddhist point of view, an example of the kind of thinking that has been common to this tradition for some twelve hundred years, and in older forms of Buddhism for twice as long. However, it is the enormously detailed knowledge that this picture encapsulates that gives contemplative traditions the potential to unravel and cure the myriad mental and emotional maladies afflicting modern life. Applied to the Western mind (which in some ways differs from the mentality of traditional Asian contemplatives), and with the collaboration of Western psychologists and neuroscientists, such knowledge and understanding could revolutionize mental health. Furthermore, this revolution has already begun. Two promising areas that are being explored are the use of enhanced awareness to deal with negative emotions, and the new field of positive psychology.[6]

STOPPING ANGER IN ITS TRACKS

A secretary cradling a pile of documents in one arm is waiting for an elevator. People rush by. One of them accidentally brushes her arm, causing the papers to spill onto the floor. As this happens, the camera closes in on her face. There a hint of irritation crosses her brow only to be erased by a knowing smile. Her eyes flash with the message "I don't have to go there." Then she laughs and stoops down to pick up the papers with a smile on her face. This vignette—from a Brazilian TV advertisement for an herbal tranquilizer—suggests that negative emotions can be attenuated as they arise. However, one doesn't need drugs.

Introspective awareness, the close monitoring of one's thoughts, emotions, and actions, is fundamental to contemplation. Training the attention in this observational skill is one of the initial steps in Buddhist and other forms of meditation. For example, hostility is considered to be one of the most virulent mental afflictions in Buddhism, a manifestation of one of the "three poisons" of hatred, craving, and delusion; therefore, to guard against its arising and expression is extremely important. For this one uses introspection. It is easily observed that in most situations where negative emotions are exhibited, they arise and play out automatically. The secretary whose documents have been scattered on the floor often shouts, "You . . . [expletive]!! Look what you did!" Fans from opposing soccer teams who begin by taunting, "We're gonna' beat you bad today!" soon find themselves in a fistfight or worse. The culprit? Negative emotions allowed to run wild.

Nowhere is our vulnerability to destructive emotions and their pointlessness more painfully obvious than in ethnic conflicts such as those seen recently in places like the Balkans, Northern Ireland, or the Middle East. It doesn't take great intelligence to recognize that these rounds of violence simply breed more and more hatred and violence, which leave thousands dead or injured with no satisfactory solution and no end in sight. Imagine the familiar scene of two opposed ethnic groups shouting at each other across a wall or barbed-wire fence. What would it take for them to recognize in that moment that the escalating anger in their minds and the increasing adrenaline and muscular tension taking over their bodies is going to lead to another pointless bloodbath? Political compromises such as reparation payments, relocation, or changing borders are not satisfactory solutions because they do not affect the hatred that is basic to the conflict. Individuals and groups dominated by negative emotions will always find reasons to be dissatisfied with their lot and to oppress some other group to improve it. Indeed, it is difficult to find cultures anywhere that are not subject to these negative feelings, if only in a latent or potential form.

Buddhist psychology and ethics suggests that recognition of the pivotal role of negative mental states and their control is the key to peace at all levels. If two groups about to engage in violence took a step toward recognizing its futility, then how might they proceed? Training in meditation holds promise, as it would allow them to observe the situation more objectively and prevent the arising of anger. There is a familiar teaching in Buddhism that says that your human enemies come and go. They may die, move elsewhere, or turn into friends through changed circumstances. Your negative emotions, however, are implacable. They stalk you mercilessly throughout your life (indeed, throughout all of your lives). *They* are your true enemies.

In the modern West, outer causes and solutions to suffering are emphasized. In Buddhism one seeks the inner sources of suffering and then trains in becoming familiar with and able to attenuate them. As we have seen, scientific proof that this is possible is coming from studies of the brain's plasticity, such as those conducted by Richard Davidson. Paul Ekman has seen similar research results, saying that he believes that some of the monks he has studied have achieved permanent positive shifts in their temperament, "increasing the gap between impulse and action." According to Ekman, "What these extraordinary people can do shows us the outer limit of what humans are capable of." Indeed, perhaps we all have the potential for acquiring these abilities.

In this regard, a new field has developed to track how emotions are reflected in the brain: affective neuroscience. It has found that thoughts, feelings, cognition, and emotions are closely interconnected. Further, it has confirmed that our emotions affect our physical health through the immune, endocrine, and autonomic nervous systems. The effects of a positive, nurturing environment are seen in how our genes are activated. Not only has the left frontal cortex been found to play a role in positive emotions, but persons with an active right frontal cortex are often dominated by negative emotions. Research has also shown that lowered activity of the amygdala (located deep within the temporal lobes) and lowered levels of cortisol (the main hormone secreted by the adrenal glands) contribute to positive emotions. The good news from such preliminary research results is that through training there is a possibility of altering our emotional temperament in a positive direction without the need for drugs or invasive procedures.[7]

ACCENTUATING THE POSITIVE

Historically in the West, the existence of negative emotions has been taken for granted. In Darwin's view, aggression served the survival of the species. Freud believed that psychotherapy could do no more than make the mentally disturbed patient ordinarily happy—that is, subject to the normal flux of negative and positive emotions. Western philosophers have viewed aggression and hostility among humans as natural and permanent. And although we are often encouraged to "think positive," we view those with a permanent positive attitude toward life—"optimists"—as a rare breed. Such exceptional people are frequently highlighted by the media in "human interest" stories. What makes them interesting is that they are so unusual, at least in Western culture.

Yet evolutionary theory has never satisfactorily explained the origin of such highly valued positive emotions as empathy and compassion. According to Buddhism, our fundamental motivation in life is the quest for happiness. Love (the wish that others have happiness and its causes) and compassion (the wish that others be spared suffering and its causes) are also considered basic to human nature. Thus, from this point of view we are essentially altruistic, contradicting the generally accepted Western view that negativity is the norm. On the contrary, according to Buddhism, negativity derives from a misunderstanding of the mind and the universe we inhabit. For Buddhism the unenlightened mind is dysfunctional. According to this view, one of the greatest errors in thinking is the belief that happiness derives essentially from one's outer circumstances.

WHY SO HAPPY?

Alan remembers from his experience of living for years in Asia that the most positive, optimistic people of his acquaintance have come from a group we would normally expect to be heavily burdened by sorrow and negativity, Tibetan refugees. The Tibetans he met in his youth in northern India and elsewhere had recently escaped from their homeland, which was overrun by the Communist Chinese army. They had left behind nearly all of their possessions, family, friends, and teachers, some of whom had been murdered or died on the treacherous journey to freedom. Many of these refugees, unaccustomed to the climate of low-lying India, perished from diseases ranging from dysentery to tuberculosis. The refugee camps were crowded and dirty. Food was often scarce. Many Tibetans survived by taking difficult jobs few Indians would accept, such as road building in higher elevations, work only accomplished by pick and shovel. Their outer conditions were extremely difficult, yet the inner attitude of the overwhelming majority was positive.

Back then one often encountered cases or heard stories of monks who had been imprisoned and tortured by the Chinese for decades. Yet some of these monks claimed that the experience had strengthened their spiritual practice, providing them with a situation where they could test and improve the depth and sincerity of their compassion by maintaining an attitude of loving-kindness toward their tormentors. Some even characterized their tenure in prison as a kind of spiritual retreat. The happiness of these refugees, so deprived in terms of their material situation, was based on inner, psychological attitudes and mental balance. Something inherent in their cultural background gave them a buoyancy that lifted them above the negativity of their desperate circumstances.

Although this kind of response to extreme adversity by Tibetans (a mostly Buddhist population) seems to contradict common sense, in fact it conforms to modern research into how our attitudes are related to inner and outer circumstances. Positive feelings based on material well-being, for instance, are fleeting. Studies have shown that lottery winners, who usually experience a flood of positive emotions in the initial period after receiving their windfall, soon revert to more or less the same pattern of emotional ups and downs they had experienced previously. Psychologist Tim Kasser, in his book *The High Price of Materialism,* concludes, "Existing scientific research on the value of materialism yields clear and consistent findings. People who are highly focused on materialistic values have lower personal well-being and psychological health than those who believe that materialistic pursuits are relatively unimportant."[8]

Treating Symptoms or Causes?

Neuroscience has become increasingly able to home in on the physical mechanisms associated with negative and positive emotions, identifying, for example, a correspondence between an overactive amygdala and right frontal lobe with the negative emotions and an active left frontal lobe with the positive ones. Given this knowledge, how should we proceed in our quest to make life more meaningful and enjoyable? Drugs such as antidepressants are common and have shown positive results (along with some disturbing side effects). Yet by focusing on the physical symptoms of mental illness are we treating the real causes? What is the most effective way to deal with these problems? If an overactive amygdala or the dysfunction of some neurotransmitter has been caused by a life experience such as a childhood trauma or a dysfunctional marriage, doesn't it make more sense to untangle the resulting negative mental and emotional patterns and replace them with more wholesome attitudes?

Buddhist psychology approaches these causes through systematic cognitive intervention. One familiarizes oneself with wholesome attitudes while developing the attentional skills needed to recognize one's afflictive thoughts and emotions in the light of a deeper understanding of the psyche from which they emanate. This can have a dramatic effect on how one reacts to stimuli. For example, when negativity arises, one catches it before the spark becomes a wildfire. It is then possible to analyze the thought or emotion, looking into its origin, appropriateness, and whether it is constructive or destructive. One can also begin to counteract these flare-ups with wholesome qualities such as patience, love, and compassion. All of this leads to developing a stable psychological basis for a healthy, happy, and positive inner life. The connection with ethics and an increased understanding of the nature of the mind make this training a far deeper and more effective approach than merely encouraging oneself with slogans to be positive.

The positive psychology of Buddhism, derived from its subtle and extensive understanding of the mind, is poised to provide a major contribution to the improvement of our mental health. Furthermore, the kind of detailed, accurate, objective feedback that can be supplied by subjects skilled in meditation is the missing link needed to complement neuroscientific research on the brain and to enrich and clarify concepts and procedures originating in Western psychology. Such collaboration would fulfill the criteria proposed by William James, which we looked at in part 1. James envisioned a science of mind incorporating both the inner subjective experience of the mind and its outer physical correlates and expressions.

Such research would proceed by coordinating three perspectives: (1) the observations of the highly trained subject (the first-person point of view—that is, the one who is having the experience) with (2) data gleaned in cooperation by skilled interviewers (the second-person view-

BORDERS.

BORDERS
BOOKS AND MUSIC
68 VETERANS MEMORIAL HIGHWAY
COMMACK, NY 11725
516-462-0569

STORE: 0179   REG: 06/94 TRAN#: 2994
SALE   04/17/2010 EMP: 01196

EMBRACING MIND
     9698218   QP T   16.95

          Subtotal   16.95
          NEW YORK 8.625%   1.46
1 Item   Total   18.41
          CASH   20.41
          Cash Change Due   2.00

04/17/2010 08:06PM

Shop online
24 hours a day
at Borders.com

BORDERS.

Returns

Returns of merchandise purchased from a Borders, Borders Express or Waldenbooks retail store will be permitted only if presented in saleable condition accompanied by the original sales receipt or Borders gift receipt within the time periods specified below. Returns accompanied by the original sales receipt must be made within 30 days of purchase and the purchase price will be refunded in the same form as the original purchase. Returns accompanied by the original Borders gift receipt must be made within 60 days of purchase and the purchase price will be refunded in the form of a return gift card.

Exchanges of opened audio books, music, videos, video games, software and electronics will be permitted subject to the same time periods and receipt requirements as above and can be made for the same item only.

Periodicals, newspapers, comic books, food and drink, digital downloads, gift cards, return gift cards, items marked "non-returnable," "final sale" or the like and out-of-print, collectible or pre-owned items cannot be returned or exchanged.

Returns and exchanges to a Borders, Borders Express or Waldenbooks retail store of merchandise purchased from Borders.com may be permitted in certain circumstances. See Borders.com for details.

point) and (3) objective physical measurements of the activity of the brain and that of other physical systems (the third-person perspective). Ideally, such an approach would yield a complete, holistic picture of what the mind is and how it functions. As the many Westerners who have flocked to meditation know, these new technologies from the East are a breath of fresh air for a troubled modern world.

A Science of Consciousness? 10
One Provocative Example from the Buddhist Tradition

NEUROSCIENTISTS commonly assume that the mind, if it exists at all, is a feature of the brain—the gray matter either *is* the mind or it alone *makes* the mind. We've seen, however, that there are major, unresolved problems with this view. Indeed it is only a belief, and one backed by little hard evidence. Cognitive scientists have failed to clarify human subjectivity in relation to its central feature, consciousness, and to convincingly equate mental phenomena with brain activity.

If consciousness is not produced solely by complex circuitry—electrochemical processes revved up to some threshold where consciousness just "happens"—where might it come from? Many contemplative traditions, Buddhism especially, would say that science has been looking in the wrong direction. We should be asking an entirely different question: Where do all phenomena (not just consciousness) come from? Their short answer to that is: all phenomena are appearances to the mind. But the "mind" spoken of here includes not only the contents we normally attribute to this word but a great deal more. For Buddhism, the mind is all-embracing. It has no existence apart from appearances, and appearances have no existence apart from the mind. Even our concepts of space-time and mass-energy have no existence apart from the mind that conceives of them. Thus, there are no absolute objects or absolute subjects. Nor is there any conflict between mind and matter, nor does one hold a dominant position over the other. We may separate them at the conventional level for the sake of discussion, but no absolute dualism separates mind and matter. Like our face and its image in a mirror, they arise in mutual dependence.

Notice that this Buddhist position is not asking us to throw away phenomena, physical or otherwise, in favor of some absolutely existing mind. The necessary coexistence of mind and physical phenomena does not mean that the latter are nonexistent or pure illusion. As we saw earlier in our survey of the Middle Way, we are being asked only to let go of our belief that phenomena must have *absolute* existence. When we do so, reality takes on a refreshing new appearance.

To most of us living in a modern, material world, that sounds pretty radical. What kind of mind or consciousness are we talking about that is somehow necessary for the existence of an objective universe? Are we back to the "mind of God," or indecipherable generalizations such as "life is but a dream"? How was such an unexpected conclusion reached? Have Asian mystics in their caves simply become lost in imagination? Or have *we*?

EXCEPTIONS WITHIN THE WESTERN TRADITION

Actually, Western science and philosophy were not always so materialistic, so worried about things being absolutely real, as they are today. In fact, the West was once quite open to the possibility of nonmaterial explanations of phenomena, especially those pertaining to human beings and our minds. Three of the founders of this tradition, Pythagoras, Socrates, and Plato, expressed such attitudes. Recall that Pythagoras believed in reincarnation, claiming to remember many of his past lives. His Pythagorean Society was the principal scientific school of Greece, and his contemplative community in southern Italy mixed religion and science. Socrates, too, saw the mind as something that continued from lifetime to lifetime, and Plato, in his *Phaedo*, opposed the then-popular notion that the soul was destroyed at death. He even described an after-death state that closely resembles the Tibetan Buddhist *bardo*, where beings' minds wander before reincarnating. Reincarnation, of course, suggests that a being is something more than the physical body and brain.

Jumping forward to the early Christian era, Origen (185–254), the most prominent Christian theologian after Paul and before Augustine, believed in contemplative practices that could guide the soul in a gradual ascent to

God from lifetime to lifetime. Like Pythagoras, he believed that for a Christian, science, religion, and one's daily activities should be fully integrated. Recall also that Augustine's four hypotheses on the origin of the human soul left open the possibility of past lives.

In similar fashion we saw how William James proposed several possibilities for the nature of consciousness and its relationship to the brain. One option was that the brain transmitted consciousness, which originated elsewhere, as a prism transmits light. He was also open to the possibility of the continuity of consciousness beyond death, saying:

> When finally a brain stops acting altogether, or decays, that special stream of consciousness which it subserved will vanish entirely from this natural world. But the sphere of being that supplied the consciousness would still be intact; and in that more real world with which, even whilst here, it was continuous, the consciousness might, in ways unknown to us, continue still.[1]

Offering a view that closely parallels the Buddhist substrate consciousness, discussed below, James speculated that consciousness might be a phenomenon different from the brain but able to interact with it. From this interaction consciousness might store aspects of the personality, including memories, and retain them after the physical death of the person.

DISPROVED OR DISREGARDED?

Although William James is greatly respected by cognitive scientists, they have largely ignored his views on the nature and origins of consciousness. The same goes for the painstaking work of Ian Stevenson, who has documented remembrances of passed lives. Thus, one might assume that the theories of James and Stevenson, along with the beliefs of Pythagoras, Socrates, Plato, Origen, and Augustine, have been discarded after thorough scientific investigation. But that is not the case. Nor is this surprising, given that science has largely shied away from investigating reincarnation or the soul. Instead, it has held fast to an unproven assumption that such nonmaterial topics must be purely speculative, the province of religion and philosophy. Here most scientists have refused to seek out or examine empirical evidence that might threaten cherished beliefs. Such a head-in-the-sand attitude has traditionally characterized religious dogmatists rather than scientists.

Even though the possibility of a nonmaterial reality, either existing in conjunction with the material or even subsuming it, was long part of the West's religious, philosophical and scientific heritage, after the Industrial Revolution this view was largely eclipsed by the materialism of science. Many of the great religious and philosophical traditions of the East, however, approached the study of the universe differently, from a first-person, experiential perspective.

A BUDDHIST VIEW OF THE ORIGINS OF CONSCIOUSNESS

The routine of a Tibetan Buddhist monk or nun living in a monastery begins early in the morning and ends late at night. Aside from chores, which fall mainly on initiates, and teaching and administrative roles performed by the more senior monastics, the day is filled with classes and periods of study and meditation. After the written Tibetan language is mastered, along with the memorization of important prayers and texts, students tackle courses in logic, debate, metaphysics, psychology, ethics, and the numerous teachings of the Buddha and later adepts. Meditation instructions are given, and in private interviews students describe their understanding and meditation experiences to the teacher, who guides them to the desired understandings and realizations.

Typically this basic education requires at least a decade. Afterward some monks and nuns go on extended solo retreats. In the traditional three-year retreat, which is a requirement in some schools of Tibetan Buddhism for becoming a "lama," a small group of students is completely isolated from outside contact for three years and three months. They follow a rigorous program of study and meditation. Some contemplatives enter even longer solo retreats, occasionally lasting a lifetime.[2]

The texts, meditation guides, and teachings used by monks, nuns, and lay contemplatives are saturated with information dealing with the mind and its nature. The typical meditation sequence instructs students to first take refuge in the Buddha, his teachings, and the Buddhist community. They are then told to give rise to *bodhichitta,* the altruistic intention to achieve enlightenment for the sake of bringing other beings to enlightenment, which relieves all suffering. The instructions that follow, which are specific to the given practice, refer frequently to wisdom as revealing what the mind is and how it operates. Even the elaborate visualization of deities, which to outsiders may seem like a form of idol worship, is essentially a teaching on the nature of the mind in its archetypal representations.

Buddhism is often described as a science of mind, and over its 2,500-year history it has elaborated several major theories of the origins of consciousness. The sources of this knowledge comprise (a) contemplative revelations by highly realized practitioners, most prominently the Buddha himself; (b) a precise system of logic; (c) a scholarly philosophical tradition that has produced an enormous, detailed, and complex literature; and (d) a rigorous, peer-reviewed educational method teaching contemplative insight. These form a complementary system of checks and balances. The Buddha had an essentially scientific attitude, counseling his students to question his words and discover their meaning for themselves. He was thoroughly averse to dogma, and his empirical stance has been the backbone of the Buddhist path.

How does Buddhism approach the mind? The first and most fundamental teaching of Buddhism is the four noble truths. They are the truth of suffering, the truth of the cause of suffering, the truth that suffering together with its inner causes can be eliminated forever, and the truth of a path that leads to the end of suffering. *Buddha found the primary cause of suffering in the mind.* Therefore, the Buddhist path to the end of suffering requires intimate knowledge of the mind. The Buddhist science of mind that exists today, with its detailed descriptions of the psyche and its techniques for personal discovery of the mind's nature, is the practical outcome of this first teaching.

To be completely freed from all the afflictions and obscurations of the mind and attain enlightenment (they are one and the same), the path requires that one develop ethics both for its own sake and as a support for stable attention and mental balance. Thereafter one must penetrate to something called *primordial consciousness*. There are, in a nutshell, three dimensions to consciousness, which one traverses in sequence on this path: (1) the examination and healing of the psyche by means of introspection and other meditative practices, (2) the attainment of the extremely stable state of meditative quiescence, and finally (3) the realization of primordial consciousness. This is a general recipe for enlightenment according to Buddhism. Here we find a science of the mind with a practical goal: liberation from suffering.

The particular Buddhist view of consciousness we will examine below employs four key terms, which I will define briefly here. (Although these terms along with their Sanskrit and Tibetan equivalents are found in other Buddhist philosophies, they are defined here as used by the Great Perfection, or Dzogchen tradition.)[3]

- ▶ The *substrate consciousness* (Sanskrit: *alayavijnana*) is the mind stream of the individual sentient being maintained throughout all of its incarnations. All of the latent and subconscious phenomena that we commonly attribute to the mind and the senses are said to emerge from the substrate consciousness.

- ▶ The *substrate* (*alaya*) is the objective, empty space of the individual's mind—blank and unthinking, yet luminous—into which all objective appearances of the physical senses and mental activity dissolve when one goes into dreamless sleep and when one dies. (The Dalai Lama has characterized the substrate as "a neutral state of mind . . . calm, placid or undisturbed . . . .")[4] The substrate is experienced by the substrate consciousness. Note that although both can be modified by the neuronal activity of the brain, the brain is not their source or location. Both are immaterial.

- ▶ *Primordial consciousness* (*jnana*) is the absolute ground state of consciousness and transcends the minds of individual sentient beings.

- ▶ The *absolute space of phenomena* (*dharmadhatu*) is the absolute ground for the appearances of external and internal space, time, matter, and consciousness. All phenomena, both "inner" and "outer," emerge from it. Therefore, there is no duality between primordial consciousness and the absolute space of phenomena (as there does exist, though in a very subtle manner, between the individual's substrate consciousness and the substrate). The unity—primordial consciousness/absolute space of phenomena—is the expanded definition of the mind mentioned at the beginning of this chapter.

THE SUBSTRATE CONSCIOUSNESS

In everyday parlance we use terms such as *person* or *individual* to indicate the "self." Delving a little deeper, we might speak of the mind or the soul, but these terms are still pretty vague. Some Christian views of the soul assert the unchanging, independent identity of the individual. Buddhism does not propose the existence of a soul in that sense. Rather than such a soul, Buddhist contemplatives have discovered, through close inspection of the mind in deep meditation, a continuum of consciousness, or "mind stream." This stream is a series of arising and passing moments of awareness. No single, unchanging identity bridging these moments is observed, and because this mind stream is influenced by various phenomena, it is interdependent rather than independent. One might be tempted to equate this mind stream, the substrate consciousness, with the individual or self, or say that each individual "has" such a mind stream. However, since no independent entity is observed gluing together the moments of awareness of the substrate consciousness, calling it "the self" is arbitrary.

Through direct experience in meditation, it has been seen repeatedly that the psyche, with its thoughts, emotions, sensations, and so forth, emerges from the substrate consciousness. The psyche is indeed conditioned by the body and the physical senses, but its source is this immaterial mind stream. This view, by the way, shares certain features with the theories of Pythagoras, Socrates, Origen, Augustine, and William James, and it is also compatible with all that is currently known about mind-brain interactions.

What kind of evidence has this Buddhist "science of consciousness" produced to back up such an assertion? And how good a "science" is it in the first place? The substrate consciousness was allegedly discovered and continually rediscovered by Buddhist contemplatives using the meditative technique called *shamatha* ("meditative quiescence"). Over millennia a detailed literature of shamatha has evolved in Buddhism, which complements an oral tradition transmitting this contemplative technique from master to student. Similar to the methods of Indian yoga, the first-person subjective experience (what the student discovers from the inside) is guided by periodic detailed interviews with a qualified meditation master. Furthermore, specific changes in behavior, perception, and outlook are expected as a result of the correct practice of shamatha. Described in greater detail below, shamatha is a systematic means of training the attention that eventually results in unprecedented degrees of mental relaxation, stability, and vividness.

Meditation guidebooks and academic treatises, many of them emanating from such prestigious institutions as Nalanda University and modern-day monastic colleges, contribute to this educational process.[5] Therefore, here we find an empirical approach that parallels experimental science. If the techniques of shamatha did not consistently lead to recognition of the substrate consciousness, it would long ago have caused a stir in the Buddhist contemplative community, and the "experiments" would have had to be thoroughly checked.

A Meeting of East and West

We can get an idea of the closeness of two empirical, experimental traditions, science and Buddhist contemplation, by comparing the two careers of one man: Matthieu Ricard, a French scientist who in 1967 traveled to India to "meet great Tibetan teachers" and eventually became a Tibetan Buddhist monk. Prior to his involvement with Buddhism, Ricard was a research scientist at the renowned Institut Pasteur in Paris. His work there and his doctorate in microbiology were mentored by two Nobel Prize winners, François Jacob and Jacques Monod. As is the case with every competent scientist, Matthieu received guidance and one-on-one feedback from a series of teachers. Their expertise, along with peer review of his published papers, provided a guarantee that his conceptual understanding of microbiology and his experimental technique were good enough for him to be accepted by his peers as a bona fide scientist.

When Ricard became attracted to Buddhism and traveled to India, he had the good fortune to find two mentors who in their own field were the equals of Jacob and Monod: Kangyur Rinpoche and Dilgo Khyentse Rinpoche.[6] Dilgo Khyentse Rinpoche (1910–91) could be considered an Einstein in the world of Tibetan Buddhism. His studies included thirteen years of nearly continual solitary meditation. He mastered the techniques and liturgy of all four major lineages of Tibetan Buddhism—no mean feat—and was a teacher of the Dalai Lama. Ricard spent twelve years as Dilgo Khyentse's personal assistant. His progress in Buddhist philosophy, ritual, and meditation was thus closely monitored by one of the great masters of the tradition. "My scientific training," says Ricard, "especially its emphasis on rigor, was perfectly compatible with Buddhist practice and the Buddhist approach to metaphysics." Today Matthieu Ricard is a highly regarded teacher of Tibetan Buddhism who has in a sense returned to science, taking an active role in the exploration of consciousness by neuroscience.

PLUMBING THE DEPTHS

The technique of shamatha (meditative quiescence) develops increasingly subtle levels of attention. Variations of it are found among the practices of a number of contemplative religious orders, including those of Christian, Sufi, Taoist, and Hindu origin. The techniques (described in greater detail in part 3) do not demand allegiance to any religious or philosophical belief system. They require no religious rituals nor that one practice devotion to some deity. Therefore, shamatha is a potential bridge between contemplative and scientific means of exploring the mind. (Shamatha is one of the meditative techniques neuroscientists are beginning to examine with sophisticated technology such as the fMRI.)

To give but a brief description, with the full achievement of shamatha —an endeavor requiring months or even years of continuous, intensive training—all involuntary, discursive thoughts are pacified. With the mind drawn inward, appearances to the senses of oneself, one's body, and the environment vanish as well. According to Tibetan contemplatives, what remains is a state of luminous, clear consciousness that is the foundation for the emergence of all appearances to an individual's mind stream, or continuum of consciousness. Like the reflections of the sky and clouds on the surface of a placid lake, so the appearances of the entire phenomenal world arise within this empty, clear, ground state of the mind, the substrate consciousness. Practicing shamatha, Buddhist contemplatives have observed a principle of the mind's operation that could be called the "conservation of consciousness." As it is with energy, (described in the first law of thermodynamics), consciousness neither emerges from nothing, nor does it vanish without trace, but rather transforms into different configurations of mental states and processes. Whereas neuroscience believes that mental phenomena originate in purely physical structures and activities, such as neurons and electrochemical processes, direct contemplative exploration reveals consciousness itself as the source of mental events.

The substrate consciousness can be likened to a stem cell in that it is undifferentiated in terms of species, gender, or personality. When an individual continuum of consciousness is conjoined with a sperm and egg, it gradually emerges as a human psyche, manifesting mental images, thoughts, perceptions, emotions, memories, and the like. Such mental events are certainly conditioned by the brain and the environment, but they ultimately emerge from and dissolve back into the substrate consciousness, which is luminous and empty. Brain processes are necessary for certain mental events to manifest in humans, but they are not sufficient by themselves to generate subjective experience. It's a joint venture.

To repeat: according to these investigations, prior forms of consciousness may transform into later forms of consciousness, but they do not change into anything else. They neither emerge from physical phenomena, such as neurons and neurotransmitters, nor do they transform into them. Mental and physical phenomena do obviously influence each other, but there is no physical mechanism by which this happens. Certainly brain injuries, for example, impair mental functioning. But there is no magical occurrence whereby electrochemical events in the brain mysteriously take on a dual nature of also being mental phenomena. From a Buddhist perspective, the notion that the mind inexplicably emerges solely from physical processes in the brain is as unlikely as a genie rising like a magical vapor from a lamp.

Just as physical processes influence the mind, mental events influence the brain and body. This is an empirical fact, whether or not any mechanism is ever discovered to explain such interactions. A mental event such as perceiving danger can cause adrenaline to flow. However, this mental event does not *transform into* that physical phenomenon—thoughts do not become adrenaline. They are, as we saw in the Middle Way philosophy, mere regularities that correlate with the event. They tend to occur in sequence.

This Buddhist view provides a fresh perspective on an impasse we encountered earlier—the so-called hard problem of demonstrating how the brain might produce subjective experience. If such mental phenomena arise from and are conserved within the substrate consciousness alone, the explanatory gap is a false problem, since according to this view a leap from physical to mental in the production of subjective experience never happens.

The substrate consciousness manifests to the individual as an inner clarity during dreamless sleep, in the last moments of life, and in deep meditative states. However, when the mind is active, the substrate consciousness is still present, taking on the forms of active processes such as thought and emotion. Yet if the consciousness of individuals operates primarily within its own sphere, what is the nature of the physical universe "outside"? Do outer phenomena have a relationship to consciousness beyond mere influence? According to the Buddhist view, dimensions of consciousness extend well beyond individual mind streams.

MIND AND SPACE

Buddhist contemplatives have discovered that the substrate consciousness transcends the specific qualities and limitations of an individual's lifetime, gender, and even species. It is claimed that once one is able to rest in and explore the substrate consciousness, access may be gained to detailed memories of past lives, some of which may not even be human. These memories are associated with the substrate consciousness itself, rather than being stored in the brain, although brain damage can obstruct access to memories as long as one is embodied as a human being. Yet if one's mind stream can reincarnate in any sentient form, this throws into question the whole notion of the "individual." After death, if the mind stream can morph from species to species, what is independent? What or who carries through? It is only the substrate consciousness, conceptualized as an ever-changing flow of awareness that stores memories as well as behavioral and psychological tendencies, which in turn influence one's subsequent embodiments. Furthermore, even this, according to Buddhism, has no absolute existence.

Buddhist contemplative science has discovered a further nuance or refinement to this level of consciousness. In the Great Perfection school of Tibetan Buddhism, a difference is drawn between the substrate consciousness just described, and the substrate that is perceived by the substrate consciousness. The inner world of "subjective" consciousness, then, is composed of the *substrate consciousness,* which is the relative ground of the psyche, whether it is active or inactive, and the *substrate,* defined as a blank, unthinking state into which all objective appearances of the physical senses and mental perception dissolve when one falls unconscious. The substrate consciousness is that which can be *conscious of* the substrate, the basis from which mental appearances of the individual emerge.

As we shall see, this novel idea that something conceived of as empty can be the source of phenomena exists both in Buddhism and in modern physics. In Buddhism, however, the outer "vacuum of space" (as conceived by astronomers and astrophysicists) is not isolated from consciousness. It is intimately related to the inner realm we've been describing.

The Original Horn of Plenty?

A parallel—and possibly a meeting point at a very deep level between physics and Buddhism—can be outlined using the scientific concept of the vacuum. Accordingly, in physics a *false, or relative, vacuum* is defined as what remains after everything possible, using current technology, is removed from a volume of space. A *true, or absolute, vacuum* consists of whatever remains after having removed everything that the laws of nature permit us to remove. Important trends in modern physics also suggest that the patterns of mass and energy that make up the physical universe may actually be configurations of "empty" space, excitations of the vacuum wherein mass resembles frozen energy. Or, to put it another way, mass and energy may be oscillations of immaterial, abstract qualities of seemingly empty space. According to the physicist Henning Genz, "Real systems are, in this sense, 'excitations of the vacuum'—much as surface waves in a pond are excitations of the pond's water .... The vacuum itself is shapeless, but it may assume specific shapes. In doing so it becomes a physical reality, a 'real world.'"[7] In other words, "everything" may arise from some ultimate "nothing."

This brings us to primordial consciousness, the ultimate level of mind that Buddhists seek to penetrate. The substrate consciousness can be compared to a *relative* vacuum. It is relatively empty, but still possesses structure and energy, characterized by such attributes as bliss (spiritual joy or rapture), luminosity (an internal radiance), and a muted sense of duality between subject and object. Primordial consciousness—characterized as the absolute ground, the most basic state of consciousness—could then be characterized as the *absolute* vacuum of consciousness.[8] Like the absolute vacuum of modern physics, it entails the lowest possible state of mental activity but the highest possible potential and degree of freedom. Furthermore, whereas the substrate consciousness is conscious of the substrate—the relative inner space or vacuum of the mind—primordial consciousness is indivisibly aware of the absolute space of all phenomena (*dharmadhatu*), which is beyond the duality of external and internal space. Out of this space emerge all the phenomena that make up all worlds of experience—the whole universe, inside and out, subjective and objective. All appearances of external and internal space, time, matter, and consciousness emerge from the dharmadhatu and consist of nothing other than configurations of this absolute or true vacuum.

Contemplative technologies more subtle and powerful than shamatha are required to discover and explore the unified experience of the dharmadhatu and primordial consciousness. To reach this level, after perfecting shamatha, one cultivates contemplative insight (Sanskrit: *vipashyana*) and then achieves full realization accessing primordial consciousness through the practices of the Great Perfection. Briefly stated, contemplative insight is a subtle questioning of the activities and nature of consciousness and the rest of the world of experience, based upon the attentional stability provided by shamatha. The practices of the Great

Perfection are even more subtle, providing direct, nondiscursive access to primordial consciousness. The experience of the substrate consciousness produces states that are nonconceptual, luminous, and blissful. However, with the penetration of primordial consciousness it is said that one experiences these as a unity that is beyond words to describe.

Again, the experience of the substrate consciousness is specific to the "individual," connected with time and causality, and occurs only when the mind is withdrawn from the senses and thoughts. Primordial consciousness, however, is a nondual state where all distinctions of subject and object, mind and matter, indeed all words and concepts are transcended. There is a total openness to all phenomena but no individual or separate "experiencer." No subjective consciousness encounters an objective reality. Rather there is a realization of the intrinsic *unity* of absolute space and primordial consciousness.

In the lore and literature of the Great Perfection, this union of absolute space and primordial consciousness is called the "single taste" of all phenomena and the "purity and equality of the whole of samsara and nirvana." Lacking all internal structure, it transcends relative space, time, mind, and matter. This parallels the conception of the absolute vacuum in physics, suggesting that the dharmadhatu of Buddhist cosmology may be similar.

ZERO, BUT NOT EMPTY

The physicist Edward Tyron has proposed the theory that the universe consists of a gigantic vacuum fluctuation whose total energy is essentially zero. How could our universe, which is so apparently full of matter and energy, cancel out to such a vacuum state? Henning Genz suggests that such a universe may have originated as a spontaneous vacuum fluctuation. "Maybe quantum mechanical fluctuations initiated not only the *stuff* our world was made of prior to inflation[9] but also space-time itself," says Genz. "Maybe the true *vacuum*, the true *nothing*, of philosophy and religion should be seen as a state wholly innocent of laws, space, and time. This state can be thought of as nothing but a collection of possibilities of what might be."[10]

If indeed all that exists emanates from a vacuum of mere possibilities,

Review of the Main Features of a Buddhist Science of Consciousness

The Three Dimensions of Consciousness

THE PSYCHE	THE SUBSTRATE CONSCIOUSNESS	PRIMORDIAL CONSCIOUSNESS
Origin: initially emerges from the substrate consciousness at or briefly following conception, then develops in dependence upon physical, mental, and social influences	*Origin*: a stream of consciousness that precedes the emergence of the psyche in this lifetime and has no identifiable beginning	*Origin*: a dimension of "unborn" consciousness that transcends time and therefore has no origin
Nature: dualistic conscious and subconscious mental processes, including the five physical senses, arising from moment to moment	*Nature*: the relative ground state of consciousness, out of which all conscious and subconscious mental and sensory states of the psyche emerge, conditioned by a wide range of influences from this and past lives	*Nature*: the ultimate ground state of consciousness, out of which all appearances of samsara and nirvana emerge and into which they subside
Character: habitually in a state of dissatisfaction, unease, and restlessness	*Character*: a relative vacuum state of consciousness, withdrawn from the five physical senses and from mental activity, in which the duality of subject and object is muted, qualified	*Character*: the absolute, boundless vacuum state of consciousness, nondual with the absolute space of phenomena, it transcends all conceptual constructs, including subject and object, existence and nonexistence, and bears the
Activity: prone to mental afflictions such as craving and hostility, but is		

also activated in ethically neutral and wholesome ways

Experiential access: introspection, which, as it is cultivated and refined, can probe more and more deeply into activities of the psyche that normally occur subconsciously

Termination: the blackout phase of the dying process, when all sensory and mental appearances vanish

by the three distinct aspects of bliss, luminosity, and nonconceptuality

Activity: while wholesome and unwholesome mental processes are deactivated in this relative and resting state of consciousness, the individual continuum that is undergoing constant change stores memories, personality traits, and karmic imprints from this and past lives

Experiential access: manifests naturally during deep sleep and the blackout phase of the dying process, and can be voluntarily accessed with the achievement of shamatha

Termination: this stream of consciousness continues on through the intermediate phase following death and into future lifetimes with no identifiable end

undifferentiated qualities of bliss, luminosity, and nonconceptuality

Activity: acts as the source of spontaneous, effortless virtues, such as wisdom, compassion, and creativity

Experiential access: manifests spontaneously as the "clear light of death" immediately following the blackout phase of the dying process, and able to be accessed through such wisdom practices as Mahamudra and the Great Perfection (explained below)

Termination: abiding in the "fourth time," beyond the threefold division of past, present, and future, it is both unborn and undying, and therefore has no end if the physics of the vacuum and the unity of primordial consciousness and dharmadhatu coincide, we have returned full circle to the miraculous. Furthermore, if the universe originated in the spontaneous fluctuation of a vacuum that was not merely physical, an absolute space of phenomena that is unified with primordial consciousness, then mind and matter emerge from the same ultimate ground. Mind, whether you call it "the mind and word of God" (Logos), "the Tao," Einstein's "superior mind,"[11] or "primordial consciousness," saturates nature. The universe is not an unthinking machine; it is intelligent. Traditionally we have defined miracles as causes and events that transcend physical explanations. However, compared with what we are suggesting here, walking on water and reading minds seem like small potatoes. What could be more miraculous than a universe that is conscious? Whether "God" created the universe in six days (the literal biblical view) or countless "billionfold worlds" exist in an endless cycle alternating between destruction and creation, perhaps a never-ending series of big bangs (a Buddhist view), the implication is that existence itself is, in the most simple and direct sense, a miracle. However, here, of course, we are confronted with the biggest "if" imaginable. What kind of research could tell us if any of this is true? The only science capable of confirming such a universe would have to fully accept, explore, and integrate both the "outer physical" and "inner conscious" aspects of existence.

According to the physicist John March-Russell, scientists currently believe that a complete understanding of the vacuum is required as a prerequisite for understanding all other aspects of phenomena and the universe. Science has yet to achieve such an understanding. Despite that warning and the implications of quantum physics, most scientists still refuse to believe that consciousness is relevant to the universe they seek to understand.

According to Buddhism, consciousness is essential to reality and the understanding of it. At the level of primordial consciousness/dharmadhatu, no distinctions between matter-energy and consciousness are found. They are undifferentiated. Yet they are accessible to the human mind. Whereas the inferences made by physicists speculating on the vacuum are conceptual and lacking direct experience, contemplatives with the ability to rest in the open clarity of primordial consciousness may perhaps experience the source of universal creation. From a commonsense point of view, that, too, would seem nothing less than miraculous.

CHALLENGES FOR A MODERN SCIENCE OF CONSCIOUSNESS

One of the most controversial areas in spirituality is the paranormal. Under the rubrics of "magic" and "witchcraft," Christianity condemned this "supernatural" realm as sinful and demonic. Following in these footsteps, science denied the existence of the supernatural and miraculous, leading inevitably to skepticism about the nature of the mind as anything other than a property of matter. Buddhist contemplatives claim that the realization of primordial consciousness, when preceded by the perfection of shamatha, reveals limitless internal resources for various kinds of extrasensory perception and paranormal abilities. Among these are remote viewing or clairvoyance, clairaudience, knowledge of others' minds, precognition, and other such skills, including the power to mentally control the four elements of earth (solidity), water (fluidity), fire (heat), and air (motility). Examples here include moving through solid objects, walking on water, mental control of fire, flying, and mentally multiplying and transforming physical objects at will.

It is important to note that these abilities have never been major aims of Buddhist practice. Encountering them can even be considered problematic in that they may seduce one from the main goal of freeing the mind of its afflictive tendencies and tapping into our full potential for wisdom and compassion. At best they can be considered signs of progress along the path to awakening. Their actual use is said to require great discretion. However, there is anecdotal evidence that they are occasionally witnessed in Buddhist communities, sometimes by large crowds of people.

There have, in fact, been some interesting scientific investigations into this area. The physicist Russell Targ has made a study of remote viewing (the ability to visually perceive things not in one's immediate surroundings) and precognition (knowledge of future events). He claims these powers exist, and has gathered data to back his assertions.[12] We have already touched on the research of R. G. Jahn of Princeton University, who has investigated the ability of the mind to influence physical objects.

The work of these scientists can be likened to the experiments of nineteenth- and early-twentieth-century physicists, whose only access to elementary particles came through manifestations that are occasionally and unpredictably produced in nature. The cultivation of deep states of meditative concentration, however, compares to the use of the modern particle accelerator, where one can study elementary particles at will under laboratory conditions. If neuroscientific studies of contemplatives make headway, more physicists may begin to accept the reality and potentials of consciousness and to study its "paranormal" aspects. Some might even accept the challenge of testing contemplative technologies personally, and thereafter conduct experiments from both the "inside" and the "outside."

THE ADVANCEMENT OF SCIENCE

Our universities and scientific institutes are replete with laboratories that study the brain and behavior. Yet if the full range of conscious experience is to be understood, these should be complemented by facilities designed for the study of refined states of consciousness. The cognitive sciences have undoubtedly made useful contributions to our knowledge of normal and subnormal mental states, much as classical physics has aided our understanding of the macro-universe. However, we have seen that many of the underlying beliefs of classical physics were never true, failing to explain areas where quantum physics and relativity have been far more successful. In like manner, the physical approach to the mind currently popular in the cognitive sciences may prove inadequate when exceptional states of consciousness are developed under controlled conditions and studied with scientific rigor. If claims from both contemplatives and scientists regarding extrasensory perception are shown to be well founded, then it becomes necessary to propose a dimension of consciousness transcending our normal views of time and space, one extending beyond the psyche and even beyond the substrate consciousness.

Among several crucial challenges facing the development of a modern contemplative science, one is to reaffirm the existence of consciousness in nature without reducing it to a secondary property of matter, a by-product of the brain. Here Cartesian dualism has proven futile as a model for understanding consciousness, and we have seen that scientific materialism contradicts personal experience and that it is based on out-dated beliefs from nineteenth-century physics, thus severely limiting our understanding of the nature and potentials of consciousness.

We must also undertake a fundamental reassessment of human nature. Viewed from the perspective of classical physics, humans are reduced to robots. Biology reduces us to animals. Contemporary, main-stream psychology has focused primarily on the study of normal and subnormal minds, thereby limiting its vision of humankind. Buddhism takes a more open and essentially optimistic view, expressed in three dimensions: (1) human nature as defined by the human body and psy-che; (2) our nature as sentient beings, based upon the individual sub-strate consciousness that can give rise to a wide variety of psyches, human and otherwise; and (3) our Buddha-nature, characterized by pri-mordial consciousness, which transcends all limitations of individual human or sentient existence.

As for our view of reality as a whole, we have argued that as long our notion of consciousness is confined to normal and subnormal mental states, the underlying principles of classical science are approximately valid for an objective physical universe devoid of subjective awareness. Neglecting the role of consciousness may seem insignificant, much as the principles of classical mechanics appear valid as long as one excludes elementary particles and matter that approaches the speed of light. However, when one includes quantum and relativistic phenomena, things change drastically. Likewise, when consciousness is highly re-fined, its relevance to the physical world becomes crucial, as we have seen in both modern physics and contemplative traditions. Here the materi-alistic assumptions of classical physics (and philosophical hangers-on such as scientific materialism) are inadequate, and we must take into account all dimensions of consciousness and no longer equate it with the psyche alone.

The scientific investigation of refined states of consciousness has the potential for revolutionizing the cognitive sciences. Furthermore, given that natural science as a whole is presently based on the assumptions of materialism, such explorations could transform science from top to bottom. The essential enterprise here entails a thorough investigation of the full scope of consciousness, in particular the role of refined states of consciousness in nature. The result may well be an all-embracing science of the world of experience to replace our current science of a purely objective, physical universe in which subjectivity is excluded.

A NEW ROLE FOR POLITICS

The contemplative traditions that once influenced Western culture have all but vanished from modern life. We may turn our heads toward the East in hopes of reviving them, but we can still tap into roots found closer to home. "It is requisite for the good of the human community that there should be persons who devote themselves to the life of contemplation," wrote Saint Thomas Aquinas, whose influence on Christianity and Western culture cannot be overestimated.[13] Might not the very purpose of civilization be the pursuit of genuine happiness, truth, and virtue, based upon a life engaged in contemplation? And wouldn't that require that this—the essence of a meaningful life—extend naturally to the sphere of politics? That, I believe, is what Aquinas had in mind when he wrote "The whole of political life seems to be ordered with a view to attaining the happiness of contemplation. For peace, which is established and preserved by virtue of political activity, places man in a position to devote himself to contemplation of the truth."[14]

The Dalai Lama has stated, "In essence, science and spirituality, though differing in their approaches, share the same end, which is the betterment of humanity." He then commented, "At its best, science is motivated by a quest for understanding to help lead us to greater flourishing and happiness . . . ; this kind of science can be described as wisdom grounded in and tempered by compassion. Similarly, spirituality is a human journey into our internal resources, with the aim of understanding who we are in the deepest sense and of discovering how to live according to the highest possible ideal. This too is the union of wisdom and compassion."[15]

During the past century, the human race has acquired more knowledge and power than in any other period in recorded history. But we remain as vulnerable as ever to the deep-rooted afflictions of arrogance, greed, and hatred and the untold suffering they cause. When these distortions of the human spirit drive our search for knowledge and power, tragedy strikes in myriad ways that imperil our well-being and, ultimately, our very existence. In the interests of our welfare, indeed our survival, the current conflicts between science and religion must cease. An uneasy truce is not enough. Nor is it enough to merely argue that science and religion are compatible, without challenging the assumptions of both and extending the scope of human experience through mutual engagement. The time has come to join hands, uniting the knowledge and power of science and technology with the wisdom and compassion that characterize genuine spiritual realization. With the abandonment of all dogmas, let us return to a true spirit of empiricism, embracing the whole of nature, one that includes the objective world of science, the mind, and spirituality, and in so doing bring about a new renaissance that will heal our troubled world.

PART THREE
Tools and Technologies of a Buddhist Science of Contemplation

WHEN WE READ any book, we embark upon a journey of the mind. On this particular trek we have examined the mind as a "knower"—from limited notions such as "seeing is believing," on through a maze of scientific and philosophical views, finally reaching lofty visions of primordial consciousness as one with the universe itself. In parts 1 and 2 we have said all we have to say about the mind in this book. Some readers, however, may be curious to know more about the potentials of meditation. For those who would like a more intimate glimpse into the contemplative technologies mentioned in chapter 10, "A Science of Consciousness," part 3 provides an overview of contemplation from a Tibetan Buddhist perspective. This includes brief explanations of the practices of *shamatha* (meditative quiescence), *vipashyana* (insight meditation), the Great Perfection, and several other relevant techniques. (As mentioned previously, shamatha is a preliminary practice for making the attention extremely stable and vivid. Based on this skill, one then explores the mind in relation to experienced phenomena, critically using insight meditation. Following the traditional sequence, one finally is able to use the more direct and nondiscursive methods of the Great Perfection.)[1] We have also included some information on ethics in Buddhism and how it is connected with meditation, daily conduct, and the overall vision of this spiritual tradition.[2]

Shamatha
Meditative Quiescence 11

To ACHIEVE mental and emotional quiescence, to be able to abide calmly in one's life, would be a welcome relief to almost anyone in any period of history. All the more so in the turbulent world we live in today. As we've seen, such states of quiescence have been cultivated in a variety of spiritual and cultural settings. Many Indian contemplatives over 2,500 years ago practiced shamatha for the sake of achieving experiences of peace that surpass conceptual understanding. Believing that such states of relative quiescence did not uproot the true causes of suffering, the Buddha taught shamatha as a tool for venturing farther. The perceptions revealed by using this exploratory tool would lead the contemplative to self-knowledge, the key to the door of enlightenment. We have argued that shamatha, when combined with contemplative inquiry, can lead as well to empirical knowledge, even a scientific understanding of the mind and certain elements of the universe at large.

Within some Buddhist traditions, shamatha is the first step in a sequence of practices. Once it has allowed one to fathom the relative ground state of consciousness (the substrate consciousness), *vipashyana* (insight meditation) is used to explore the nature of reality as a whole. Thereafter the Great Perfection can facilitate the penetration of primordial consciousness. Each of the four major schools of Tibetan Buddhism has its own techniques for accessing primordial consciousness. They vary in some ways, but they also bear a number of fundamental characteristics in common. They are as follows: the Great Perfection (Dzogchen) from the Nyingma order; Highest Yoga Tantra from the Gelug; the Union of Emptiness and Clarity from the Sakya; and the Great Seal

(Mahamudra) from the Kagyu school. Strictly speaking, the Great Perfection is a part of the Vajrayana ("The Diamond Vehicle") and is traditionally approached after having achieved some degree of realization in the more basic practices of Vajrayana. (The techniques of Vajrayana, also called Mantrayana or Secret Mantra, are incredibly vast and include a wide variety of visualizations, sometimes connected with the manipulation of psycho-physical elements and the breath. The term *tantra* is also connected with these practices, although strictly speaking, the word refers specifically to Vajrayana texts.)

For attaining spiritual goals, motivation and ethics are critical for practices originating in the Mahayana ("Great Vehicle") category of Tibetan Buddhism, which includes Vajrayana and the Great Perfection. Here practice is to be motivated by *bodhichitta,* the wish to achieve perfect enlightenment in order to free all beings from the sufferings of the cycle of birth, aging, sickness, death, and rebirth. Ethics, as mentioned earlier, is important both for its own sake (especially in connection with bodhichitta) and as a lifestyle support contributing to a stable mind.[1]

Techniques such as dream yoga (lucid dreaming) and *tummo* (a complex psycho-physical practice that employs body heat) are used to enhance one's progress. Note that the sequence presented here—shamatha-vipashyana-Great Perfection—is by no means the only one used in Tibetan Buddhism, and details within this particular progression are usually decided upon according to the meditation teacher's evaluation of the student. The descriptions below are given merely to provide an overview of the techniques of this contemplative science and are not appropriate as instructions upon which to base one's practice. That is best provided by a meditation master or at least through the study of books written by bona fide instructors.

Shamatha practice is traditionally divided into nine stages, after which one is said to have "attained shamatha." The sequence we present below derives from the classic text *Stages of Meditation* by the Indian adept Kamalashila. He was one of the earliest Indian *mahasiddhas* (highly accomplished yogis) to teach in Tibet, invited there in the middle of the eighth century by King Trisong Detsen (790–844).

These nine stages fall into three general sections. In the first (stages 1–4), the student concentrates on a meditative object, such as the tactile sensations of the breath, though sometimes visualizations are used. The goal here is to use attention focused on a meditative object to overcome the gross level of distraction—being carried away by daydreams, sense data, strong emotions, mental chatter, and the like. In the second section (stages 5–7), awareness is gradually withdrawn from the five outer senses to the sixth sense and its object—mental phenomena. Here, more subtle distractions are overcome by allowing the attention to rest at ease while mental phenomena pass through undisturbed. In section three (stages 8–9), the most subtle remaining distractions are eliminated as one withdraws the attention from all objects, eventually allowing awareness to rest in its natural ground, the substrate consciousness.

Attaining shamatha can be a challenging and lengthy endeavor. The basic qualities that lead one forward are stable and vivid attention. "Stability" refers to maintaining concentration steadily on the chosen object—one does not stray. "Vividness" refers to the degree of clarity or closeness of attention one maintains on the object. To develop and maintain these qualities, mindfulness (remembering the task) and introspection (checking, quality control) are needed to monitor one's moment-to-moment experience. This must be done in order to avoid the twin obstacles of agitation and dullness. According to Buddhist psychology, a moment of consciousness lasts only milliseconds, and these brief pulses of awareness are more and more distinctly ascertained as one advances in this practice. Verifying this possibility scientifically would go a long way toward vindicating William James's belief in the potentials of introspection for the exploration of consciousness.

How long does it take to achieve shamatha? The Tibetan oral tradition says this depends on the talent of the student as well as on having the right conditions—having an experienced guide, living in a conducive environment with few distractions, being well prepared, and practicing diligently. Those with "sharp faculties" may achieve shamatha in three months, those with "medium faculties" in six months, and a person of "dull faculties" may take nine months. However, there is a significant difference between the context of Tibetan hermits and monastics living in the isolated reaches and relative tranquillity of Tibet prior to the Chinese occupation, and those of us in the modern West. For us I would agree with modern psychologists, who calculate that five thousand hours of concentrated training—eight hours each day for fifty weeks in the year, for a total of nearly two years—are generally required for the accomplishment of many high-level skills. This gives a rough idea of the level of commitment necessary for attaining shamatha.

THE NINE STAGES OF SHAMATHA

Stage 1. Directed Attention

The untrained mind is often distracted. This is due in part to the fact that we live in a distracting environment, our inner distractions often reflecting those of daily life and vice versa. Decades of this kind of mental life make distraction a strong habit—the norm. We can arrive at a state of mental quiescence by concentrating on just about any object. The basic idea is to ignore the distractions coming from the six senses and instead guide our attention back to the object of concentration whenever it wanders. In order to do so, we use mindfulness and intro- spection. We must notice that we have become distracted. If we perse- vere, over time the habit of concentration begins to gain strength as the habit of distraction subsides.

In the initial stage the attention moves back and forth between focus on the meditational object and being distracted from it. Concentration on the breath—practiced and recommended by the Buddha himself— is one of the most common methods. The student may concentrate on the sensation of the rising and falling of the abdomen or on the flow of breath through the nostrils. The Buddha admonished practitioners to attain full concentration on the in- and out-breaths, being aware of the length and sensation of the entire cycle of breathing: "Breathing in long, one knows, 'I breathe in long.' Breathing out long, one knows, 'I breathe out long.' Breathing in short, one knows, 'I breathe in short.' Breathing out short, one knows, 'I breathe out short.'" In addition we can count the breaths, which acts as a further check on distraction.[2]

At first it is common to observe that we seem less concentrated and more distracted than usual. What is actually happening, though, is that we are becoming aware of just how noisy and unstable our minds nor- mally are. The seeming chaos of distraction is our ordinary mode of operation, only now we are tuning in to it, which makes it seem all the louder. So we enter a process of focusing, then becoming distracted, noticing our distraction, and bringing our attention back to the object of concentration. We discover how easy it is to become lost. We notice that the air coming in through our nostrils is colder than the exhalation. We are reminded of the cold wind sweeping in off the lake or the ocean in winter. We remember snow blowing outside the window at Christmas and the colors of the tree ornaments and smells of cookies baking in the oven. Mother is taking the cookies from the oven. Mother is wearing a bright yellow apron. It's just like the bright yellow Ford Mustang of a former girlfriend. We become annoyed remembering how she infuriated us by running off with a friend. Perhaps our agitation then awakens us from this dream, a few seconds but light-years distant from our original intention of concentrating solely on the breath. This is how it is at the beginning. With persistence we gain stability, and gradually these flights of fantasy diminish.

Before continuing on through the stages of shamatha, we must mention another form of meditation that plays an important role here. *Discursive meditation* makes use of the verbalized thought activity that is normally avoided in meditation. When we deliberate mentally in a very clear and orderly manner on any topic, that is discursive meditation. Since we are going against the normal flow of our mental habits in practicing shamatha, it is useful, at least in the beginning, to verbalize what we intend to do, how we are going to go about it and even to offer ourselves words of encouragement. This is usually done at the start of a session as a way of centering and orienting oneself.

Stage 2. Continuous attention

We have reached this second stage when we are occasionally able to maintain stable concentration on our chosen object for around a minute. Still, during most of the meditation session we are swept away, distracted by thoughts, emotions, and sensations.

Stage 3. Resurgent attention

Here the increasing power of mindfulness begins to tell. We are able to spend the majority of our session concentrated to some degree on the object of attention. On occasion we still lapse into distraction, but we quickly become aware that we've strayed and we bring ourself back to center.

Stage 4. Close attention

At this point our enhanced mindfulness will not permit us to completely forget the object of attention. Though our clarity may not always be strong, we still maintain our attention consistently on target. This signals freedom from coarse excitation. Coarse laxity and medium excitation, however, are still present. In order to move on to the next stage, it is generally considered necessary to shift from a sensory object to a more subtle, mental object of attention.

Stage 5. Tamed attention

Sometimes at this stage in the practice of mindfulness of breathing, a particular mental image spontaneously appears to us. In that case we are instructed to maintain our attention right there. In this fifth stage, we may also be instructed to "rest the mind in the natural state." Here, rather than resisting the intrusion of mental phenomena as done previously, we allow thoughts, emotions, images, and such to arise and pass without any interference. Gradually the attention becomes so relaxed that mental phenomena can be observed closely without the occurrence of any grasping (attachment to the phenomena) or distraction. We dwell exclusively in the present, with memories of the past or anticipation of the future unable to nudge us off center. The power of mindfulness is now becoming honed to the degree that we can detect relatively subtle occurrences of excitation and laxity.

Stage 6. Pacified attention

Through the increased power of mindfulness we no longer feel any resistance to shamatha training. Our main obstacles now are medium laxity and subtle excitation. The object of attention may not be perfectly vivid, and very subtle, distracting thoughts may keep us from being perfectly on center.

Generally speaking, the practice of shamatha can be likened to listening to music. Imagine you are at a party and a melody playing on the stereo catches your attention. You wish to follow it closely, but another guest may pull you aside and ask you about a mutual friend, and before long you are focused on that. You may then drag yourself away from the guest or ignore him and refocus on the melody, reestablishing your "stability." Then someone tells a long, boring story, and your attention withdraws and succumbs to laxity or dullness. Here the "vividness" of your attention has been diminished. Yet had you taken a close interest in the person telling the story, even though the narrative itself was uninteresting, you could have maintained stable and vivid attention no matter what.

Note that by this sixth stage thoughts are so infrequent that our reference points informing us that we are an individual begin to fade. This may be frightening. Also, in this relative absence of thoughts, it is possible that repressed memories, fantasies, and emotions, some of which may be deeply disturbing, could surface. These vary according to the individual. We are counseled not to become attached to them, but rather to simply observe their arising and passing.

Stage 7. Fully pacified attention

Now an increased level of enthusiasm makes the practice joyous and uplifting. Attachment has been reduced to the point that the few waves of involuntary thoughts or emotions that appear roll through undisturbed. Only a very slight slackness of attention—subtle laxity—remains. And subtle excitation occurs only occasionally such that we may meditate up to several hours without the least disturbance from either excitation or dullness. The Indian adept Asanga described it in this way: "Attachment, melancholy, and so on are pacified as they arise." At this point we may also practice *quiescence without a sign*, which means maintaining stability without any object whatsoever. Awareness rests naturally in its luminous, cognizant nature. This is a preparation for the advanced contemplative practices of vipashyana and the Great Perfection.

Stage 8. Single-pointed attention

Here the overall quality of attention is stillness and our attention has become so unified that it can be directed at will. Even the most gentle and subtle distractions occasionally encountered in the previous stage are absent. Our experience is likened to the unruffled calm of an ocean free of waves.

Stage 9. Attentional balance

Very little effort is required to move from the previous stage to this ninth level. Here continuity can be maintained flawlessly for at least four hours. This is the level that immediately precedes the attainment of shamatha, and we have gained the ability to enter into it effortlessly and at will. On the actual attainment of shamatha it is said that we experience a sensation like that of a hand being placed upon our shaven head. This is associated with our having gained freedom from mental dysfunction. Our attention has become supple and powerful, a precision instrument entirely at our service. Furthermore, tremendous energy is felt flooding through our body. This feeling signals a shift in the operation of our nervous system and an end to physical dysfunction. A light, vibrant quality pervades the body. These experiences of mental and physical bliss gradually subside, but perfect stability and vividness of attention remain as long as we do not let our practice deteriorate.

It may be clearer now why we earlier compared the development of shamatha with the creation of the telescope. Like today's advanced telescopes, shamatha provides the viewer with a stable base, allowing one to sweep the heavens without fear of losing one's way, and a powerful lens bringing distant and otherwise invisible objects into sharp focus. This is precisely the kind of instrument needed to successfully explore the immense, complex, and dynamic universe of the mind.

Vipashyana
Insight Meditation 12

S O POWERFUL and liberating is the attainment of shamatha that it is sometimes mistaken for enlightenment. However, in Buddhism shamatha is called merely "access to the first stabilization (*dhyana*)." Within that context it is the first major step in a series of contemplative practices leading to enlightenment. Along the path to full awakening shamatha provides only temporary relief from the symptoms of mental dysfunction. The more advanced practices that follow are based upon shamatha and provide the cure. For those interested in the scientific, psychological, and philosophical applications of these practices, shamatha can be said to provide access to the inner workings of the mind. Then vipashyana, which is more subtle, allows one to cut through misleading appearances and fathom the actual nature of phenomena. By pointing out our superimpositions—the projection of subjective views that distort our perception—vipashyana allows us to experience reality as it is.

From a Buddhist standpoint our mental afflictions, or distortions, stand in the way of enlightenment. From an empirical or scientific standpoint, such biases impede the search for truth, especially since the mind is truly the primary scientific instrument. Whether we are trying to use the mind and scientific instruments to probe stars and galaxies or we wish to understand the nature and workings of the mind itself, our mental projections and illusions of knowledge cloud the picture.

There are a variety of approaches to vipashyana. The four applications of mindfulness, explained below, comprise a classic method, originally taught by the Buddha. Although our explanation will be primarily in terms of spiritual aims, the common goal of both religion and science—seeking the truth—should make the application of these techniques to science and philosophy clear.

The Four Applications of Mindfulness

The Mindfulness of Body

"In the seen there is only the seen; in the heard,
there is only the heard; in the sensed there is only the sensed;
in the cognized there is only the cognized."

These are the words the Buddha used to characterize the practice of vipashyana. According to Buddhism, the sources of ignorance are both a passive dullness, or ignorance, and an active misconstruing of reality. The quote above offers an antidote to such conceptual projections. Instead of seeing "only the seen," we normally project distortions, based on our dysfunctional minds, upon reality. In the more or less passive practice of shamatha, we clean the lens of our mental perception. With vipashyana we use that lens as part of an active process of inquiry into the nature of reality. The questions that guide our observations reveal just how subtle and pervasive our projections can be. They range from personal biases up to cultural beliefs and assumptions. Suppose, for example, that through the practice of vipashyana, Albert Einstein had noticed that his adherence to the classical notion of locality (the basis of the flawed EPR paradox) was rooted in a psychological need for a secure, easily visualized understanding of reality. Of course we are merely speculating, but such a revelation might have set him to thinking, "Is locality a feature of the physical universe or just something I have been projecting upon it?" This is the kind of question asked in vipashyana. Who knows, had Einstein asked it of himself, he might have arrived at Bell's inequalities thirty years before Bell did.

During the practice of the mindfulness of body we experientially examine our own body, investigating the factors of origination, presence, and dissolution of bodily processes. If we are directed to feel the top of our head, we normally include a mental image of our head, which often then becomes mixed with the pure bodily sensation. We are not perceiving "only the sensed" but superimposing something extra. By using our full attention and repeating this practice over weeks or months, we begin to discover all kinds of subtle interferences our minds place between us and our sensations. During practice sessions, mindfulness is also directed toward the visual, olfactory, aural, and gustatory senses. We taste something in our mouth, but are we able to taste it directly or is the thought "Yuck! That tastes bad" part of the experience? We discover that the way we experience things at very subtle levels is often influenced by subjective habits. Smells are either "good" or "bad"; sounds are either "loud" or "soft"; sights "bright" or "dark." In this practice, however, we are trying to perceive with discerning mindfulness, unclouded by preconceptions.

The Mindfulness of Feelings

Here we investigate the nature of feelings related to sensory and mental experience, specifically the feelings of pleasure, displeasure, and indifference. (In Buddhism feelings that are neither pleasant nor unpleasant are still considered feelings. For example, apathy with its neutral tone is still a feeling.) We ask ourselves such question as "How stable are my feelings?" "What causes them to change?" "Does observation of my feelings cause them to change?"

Our feelings influence everything we do. They are often responsible for our misunderstanding reality, our wrong conclusions leading to wrong decisions. In Buddhism, slavery to our feelings is summed up in the eight mundane concerns, touched upon earlier: We oscillate helplessly between the desire for material gain and the avoidance of loss, of pleasure and pain, praise and criticism, fame and ridicule. By fathoming the nature of our feelings, we give ourselves time to gauge our responses in day-to-day activities.

Suppose it is our habit to become angry when a family member forgets to lock the doors to the house at night. It may be that although we now live in a safe neighborhood, we were ourselves severely scolded in childhood for committing this same offense. Through the mindfulness of feelings we may discover the nature and origins of this phobia. If so, the next time our son or daughter leaves the door unlocked at night we will be able to be sensitive to the very first hints of rising anger and to determine beforehand if the ensuing argument is worth it or not. By thoroughly investigating our feelings we take a giant step toward freeing ourselves from their dominance.

The Mindfulness of Mind

Attention is now focused on the sixth, or mental sense, and the mental phenomena it reveals. By way of the vividness and stability of shamatha, we now have the ability to allow thoughts, feelings, images, fantasies, memories, desires, and so forth to appear without interference. Among the experiments we can perform here is to create a single thought and observe its arising, persistence, and disappearance. From where does it arise? Where does it stay? Where does it go? How is a thought similar or different from the "mind" to which it appears?

What's the "weather" like in our mind? Is it calm, slightly agitated, blustery, tempestuous? How often does the weather change? Why? Do we identify with our various mental phenomena? Are those thoughts, feelings, images, and so forth "ours"? Which ones do we identify with? Which ones not? Why? Do we have any degree of control over these mental phenomena? Suppose we were to give up all control and just let them "do their own thing." "Who" gives up control? What part of the mind is "oneself"?

Beyond the knowledge that can be gained here on the contents and operations of the psyche, this practice leads to knowledge of self-identity, the realm of the ego. This is depth psychology par excellence. By questioning mental phenomena as they appear, the veils that cover successively subtle levels of the psyche drop, revealing the secrets of the mind and self. Shamatha has given us access to the substrate consciousness from which all subjective mental events arise. Vipashyana allows us to test the veracity of the Buddha's declaration that the inherently existent self is illusory.

The Mindfulness of Phenomena

Finally this inquiry is thrown open to all phenomena—the entire universe of experience. Here we have the opportunity to verify the interdependent nature of reality. Are so-called outer, objective phenomena truly independent of the mind that perceives them and conceives of them? By closely examining our modes of thought and perception in action, we can view the ways we grasp or interpret the objects of experience. Having deconstructed our superimposed views of reality, we have a clearer notion of how reality is constructed. This throws into question the whole notion of the "objective reality" assumed to exist by common sense and classical physics. What does "discerning mindfulness" reveal?

One of the Buddha's insights is the extent to which verbally designating, or naming things, has power to shape our experience of reality. "Nameless are all conditions," he said, "but illuminated by name." Does naming something, in the sense of conceptualizing it, bring it into existence? When we examine closely, we may discover that conceptualization goes much deeper than we normally assume. How do we separate an oak table from the tree, acorn, soil, sunlight, and water that preceded it or the rotting boards or ashes to follow? How do we distinguish between the subatomic particle and the accelerator where it briefly manifests, the minds that designed the accelerator, the theories that may have conceptualized all of these into existence, and the mental projections of the person "observing" the event? If such events occur in time, and if time is something we project onto change, what are events? If the true existence of events or objects is questionable, what is it that's changing? These crucial questions impact equally on mind, matter, and spirituality.

If conditions are illuminated by naming, what is it that does the naming? If it is the mind that names, what is that? Is it separated from the conditions—from outer phenomena—by skin and bone? The mind names the self: "I am Joan." The mind names objects: "That is Neptune." Do these locations—outside and inside—have any reality apart from mere convention? What is the nature of separation? Is the space of awareness separate from "outer space"? In the practice of vipashyana, the pursuit of answers to these questions is not merely philosophical or intellectual, it is experiential.

Viewed in terms of our projections, we make three major mistakes that can be verified through the practice of vipashyana:

1. We believe things are permanent when in fact they are in constant flux. As in our example of the "table," we tend to emphasize the reality and continuity of an arbitrary label, forgetting that our experience

of the "table" is strongly influenced by our past memories, language, and concepts, which structure our everyday experience of reality.

2. We mistake the true sources of happiness and instead pursue facsimiles that lead us to suffering. This is especially true when we seek gratification from material objects. We desire the new car, home, or gadget because we have projected qualities onto it, attributes we assume will contribute to lasting happiness. However, such happiness is always fleeting and ultimately we feel let down. Yet we rarely look inward to examine the process by which we are left unsatisfied. If we did, our newly acquired self-knowledge might lead us in the direction of greater happiness, whose source lies within.

3. We attach to certain phenomena the feelings and conceptions of "I" and "mine," when in fact there is nothing in them that is "I" or "mine." (Recall the wristwatch example where we saw that "our wristwatch" contains nothing in itself that makes it "ours." We merely project an attitude of possession upon it.)

All of this is done through the habit of grasping or attachment. According to Buddhism and other contemplative traditions, by examining and deconstructing these illusions at a very deep and subtle level, we can free ourselves psychologically and experience the awareness that precedes the demarcations of "outer" and "inner."[1]

The Great Perfection 13

IT IS OFTEN SAID that the Romans used the method of 'divide and conquer' to rule their vast empire, and in a similar way we all normally use that method to comprehend our world. In order to determine what something is and how it operates, we look to distinguish it from other things and their operations. This is duality, or dualistic thinking. To conceive of things otherwise is unimaginable, and that is just what the Great Perfection is—unimaginable. The realization of primordial consciousness, which is the ultimate goal of this practice and integral to enlightenment, is said to be beyond description. It is not a philosophical view but a direct recognition of reality. There is no duality between the recognizer and the recognized, nor between some inner space of the mind and outer space of objective phenomena. As suggested in chapter 10, primordial consciousness and the dharmadhatu have the same nature. The practice of the Great Perfection leads to the verification of this truth.

Explanations of the practices of the Great Perfection can appear misleadingly simple, often luring seekers far astray from any authentic path. This is one of the reasons these practices were traditionally kept very secret in Tibet. In an old Buddhist story, a group of blind men were each asked to describe an elephant. Each touched a different part of the elephant's body and each came up with a very partial and biased description: "like a rope," "like a pillar," and so forth. Lacking vision, none had a chance of accurately describing such an animal. The risk is similar in trying to describe the Great Perfection.

We can categorize contemplative techniques as tending toward either development or discovery. The Great Perfection fits into the latter. The developmental approach views enlightenment—the full realization of one's Buddha-nature (a synonym for primordial consciousness)—as an arduous task whereby one corrects one's lifestyle, purifies the mind, and eventually arrives at primordial consciousness. This is a gradual path of attaining what is believed to be potential in the mind. In the discovery approach, it is assumed that one's Buddha-nature is already fully present; it is perfect from the beginning. The mind's "faults" are misapprehensions of its own innate perfection. Nothing is to be rejected or, for that matter, accepted. We realize the "great perfection" of everything. In shamatha, for the most part, we are using the will, mindfulness, and introspection to develop certain qualities. Only occasionally do we release and allow for discovery. In vipashyana we let go more frequently. However, in the Great Perfection, release and discovery are the heart of the practice.

Therefore, Great Perfection practice essentially asks us to let go of our conditioning, our deeply ingrained distorted views and assumptions. We are asked to do *less* rather than more, in fact to do *nothing at all*. The instructions sound simple and poetic: "Rest in unstructured awareness." "Let go of appearances." "Don't grasp onto memories of the past or anticipations of the future. Do not dwell on the present." To be effective, however, this form of meditation must be practiced within the supportive context of the Great Perfection view and way of life, a point overlooked by many enthusiasts of this tradition today. Ideally one is neither meditating (in the purposeful sense) nor distracted, but rather a stable openness is maintained. Those who can get used to that and can sustain it until it is the way they *are*, realize enlightenment.

This sounds easy until we realize just how deeply we are conditioned and how profoundly our view of reality and way of life must shift in order for such meditation to be effective in revealing our innermost nature. Doing "nothing at all" turns out to be one of the most difficult things to do. Therefore, from the traditional viewpoint, Great Perfection practice is unlikely to lead one to the penetration of primordial consciousness and full enlightenment if practices such as shamatha, vipashyana, tantra, and so forth have not effectively prepared one.

Returning to a theoretical description, the realization of primordial

consciousness is intimately connected with the recognition that all phenomena are empty of inherent, independent existence as described in
the Middle Way philosophy. When this is realized deeply and directly, far
beyond a merely intellectual or conceptual understanding, knowing the
emptiness of all phenomena also means seeing that they are displays of
primordial consciousness. Although they are empty of inherent existence, they exist as "ornaments of primordial purity."[1] These "ornaments
of primordial purity" cannot be considered inherently existent, however,
as this view is beyond either existence or nonexistence. Or perhaps, in
the words of the physicist Edward Tyron, phenomena are "fluctuations
of the vacuum." Again, it is one thing to arrive at such an idea through
the creative imagination informed by scientific theory and quite another
to see it directly in phenomena themselves.

As we saw previously, both the Middle Way philosophy and the conclusions of William James suggest that mind and matter are conceptual
constructs. The foundations of modern physics lend support to that
view. The interdependent nature of perception, appearances, and the
concepts that describe them can be observed directly. Such an observation takes a neutral position between mind and matter: When one rests
in primordial consciousness, the nonduality of so-called physical phenomena and consciousness may be witnessed. That, however, may not
be provable in a scientific sense. The acquisition of that level of awareness appears to be an individual matter—at least for now.

"Complementary" Practices 14

HERE WE WILL briefly survey the development of *bodhichitta* as well as the nature and function of ethics within the context of Tibetan Buddhism. The aspiration to achieve enlightenment for the sake of all beings, which is one of the definitions of bodhichitta, might be irrelevant to someone using shamatha to enhance scientific understanding. However, for a Mahayana Buddhist, bodhichitta is the proper motive for entering into any activity. We have placed quotation marks around the word *complementary* in the title of this chapter because within the spiritual context these practices are essential and fully integrated with the quest leading to enlightenment.

Bodhichitta—the spirit of awakening—is at its deepest level equated with primordial consciousness. It is said that when one experiences primordial purity, the wish to also bring all beings to this state beyond suffering wells up spontaneously. In Tibetan Buddhism there exists a wide range of discursive meditations aimed at developing this powerful form of altruism. The practice of *the four immeasurables* includes a systematic examination of the sufferings and potentials for happiness of all sentient beings. The first of these, which relates to the expression of love, states, "May all beings, including myself, have happiness and the causes of happiness." Using our imagination and empathy, we visualize the kinds of happiness sentient beings seek up to and including the cause of ultimate happiness—enlightenment. The second immeasurable—compassion—is the aspiration "May all beings, including myself, be free from suffering and the causes of suffering." We imagine the myriad sufferings of beings in a wide range of realms and wish for their release.

Next comes "May all sentient beings never be separated from the happiness free of suffering." This aspiration is an expression of "empathetic joy," in which we take delight in the joys, successes, and virtues of others. The final immeasurable is the aspiration "May all beings rest in equanimity, free of attachment to those who are close and aversion to those who are distant." Here we develop the wish that all beings feel and express love, compassion, and empathetic joy toward all others without exception.

In the practice of *tonglen* ("giving and receiving"), we imagine the sufferings of other beings and take those burdens upon ourselves. We project the image of one being or a group of beings in great suffering in front of us. With the in-breath we draw into our bodies the suffering of others, often visualizing this in the form of smoke. On the out-breath, light, representing all of our positive qualities, suffuses the image of suffering beings, curing them of illness and suffering. So we are trading places with those in suffering, accepting their pain and offering them our own happiness and freedom from suffering. Using the power of empathy and imagination gained through repetition and sincerity creates a very powerful practice. We are going diametrically opposite to the normal tendency to shield ourselves from suffering while seeking to acquire all good things for ourselves. We strive to cherish others above and beyond our self-cherishing. When repeated over time and with skill, practices such as these begin to influence our overall behavior, diminishing habitual selfishness and making us more attentive to the needs of others. This is the sphere of "relative bodhichitta," and its cultivation is said to help us realize "ultimate bodhichitta," which is, as we've said, identical to primordial consciousness.

Bodhichitta is a key component in the overall development of ethics within Mahayana Buddhism. One example out of many ethical guidelines is "the ten nonvirtues." These ten activities to be avoided (with their opposites to be encouraged) can be divided into three areas: There are three nonvirtues of *body:* killing, stealing, and sexual misconduct; four nonvirtues of *speech:* lying, harsh speech, slander, and idle chatter; and three nonvirtues of the *mind:* covetousness, malice, and holding wrong views (views incompatible with reality). The student contemplates such guidelines discursively and uses mindfulness during daily activities to put them into practice. One can see that the avoidance of the ten non-virtues would naturally contribute to the practice of the four immeasurables. All of the ten nonvirtues are activities that have a negative impact, directly or indirectly, on oneself and others. By refraining from malicious thoughts about others, for example, the student is in a better position to consider the adverse effects of malice and arrive at greater empathy and compassion toward others. Similar interconnections exist between the many other ethical guidelines and the aims of bodhichitta. As mentioned previously, such ethical practices both benefit others and help in the development of a serene lifestyle conducive to contemplation and the pursuit of genuine happiness.

Whether you call it a science or an art, meditation, within its traditional frameworks, is far more than a method for achieving mental relaxation or enhanced performance. Given the many qualities we can ascribe to the mind, it is no wonder that such subtle and intricate techniques were evolved for its exploration. Shamatha alone is a highly sophisticated system of practice that can lead to extraordinary mental and even physical states, experiences so powerful that one may be deceived into believing that full enlightenment has been attained. It appears that some of the contemplative schools in India at the time of the Buddha may have been satisfied to stop there. However, for those seeking full enlightenment as defined by the Buddhist path, the quest must extend beyond shamatha.

We saw that the Buddha was not satisfied with the techniques available in the India of the fifth century B.C.E. He was seeking a final end to suffering, and the effects of shamatha meditation were only temporary. Throughout his forty-five years as a teacher, the Buddha strongly emphasized vipashyana (insight meditation) following the achievement of shamatha. To achieve a permanent transformation of the mind, an intelligent, mindful investigation into the central features of our existence is required. Investigating our own perceptions and conceptions of the body, feelings, mind, and phenomena (the four applications of mindfulness) helps uproot the bias and clinging that hold us in ignorance of our personal reality and the nature of all phenomena.

The Buddha also emphasized ethics as the primary foundation, followed by the cultivation of samadhi, or mental balance, for the exploration of the mind and phenomena. He made it clear that it is folly to tread this spiritual path if one is a prisoner of nonvirtue. First of all, stable attention is difficult to maintain if the mind is continually distracted by anger, lust, pride, guilt over misdeeds, jealousy, and so forth. These passions are so disturbing that it would be difficult even to maintain the discipline necessary to seriously enter into the practice of shamatha, and so much the more with the more advanced practices.

However, ethics is not just a method for avoiding distractions. Since it is largely based on empathy with our fellow creatures, ethics leads us in the direction of bodhichitta. The initial steps we take in considering the consequences of our actions on ourselves and others can bring us to a compassionate attitude that may blossom in the realization of primordial consciousness. Love is indeed connected to the highest levels of wisdom.

What is the place of the other techniques we have earlier mentioned only in passing, the tantric visualizations of the Vajrayana and practices such as *tummo* and dream yoga? And what of prayers, aspirations, and the recitation of mantras? The blending of all of these methods is to be directed by a person who has a deep understanding of them and a degree of mastery or realization appropriate to guide the student. The meditation master, or guru, is just as essential for modern students of Buddhist meditation as the Buddha was for his disciples.

Within Tibetan Buddhism, the Great Perfection is considered by many to be the highest, subtlest, and most direct path to primordial consciousness for those qualified to attempt it. Longchen Rabjam (1308–63), one of the practice's greatest masters, described it as the view from the highest peak, which is thus able to incorporate other views from the foothills below. As such it fully integrates compassion—the primary ingredient of the Mahayana—with shamatha, vipashyana, and the accomplishments of the Vajrayana. Viewed then in its entirety, the Buddhist path is both a science and an art.

Notes

PREFACE

1. Henry David Thoreau, *Walden* (New York: W. W. Norton & Company, Inc., 1951), 105.
2. Werner Heisenberg, *Physics and Philosophy: The Revolution in Modern Science* (New York: Harper and Row, 1962), 58.

CHAPTER 1: A SHOTGUN WEDDING

1. *Webster's Ninth New Collegiate Dictionary*, 3rd ed., s.v. "Scientific method."
2. Richard Feynman, *The Character of Physical Law* (Cambridge, Mass.: MIT Press, 1983), 158.
3. Galileo, *Il saggiatore (The Assayer)*, 1623, quoted in *Discoveries and Opinions of Galileo*, trans., Stillman Drake (New York: Knopf, 1957), 237–238.
4. *Encyclopedia Britannica Online*, s.v. "Huxley, T. H," http://www.britannica.com (accessed June 11, 2007).

CHAPTER 2: WHEN YOU LEAST EXPECT IT

1. Richard P. Feynman, R. B. Leighton, and M. Sands, *The Feynman Lectures on Physics* (Reading, Mass.: Addison-Wesley, 1963), 1–9.
2. *Encyclopedia Britannica Online*, s.v. "Atom," http://www.britannica.com (accessed June 11, 2007).
3. Cited in George Greenstein and Arthur G. Zajonc, *The Quantum Challenge* (Sudbury, Mass.: Jones and Bartlett, 1997), 84.
4. Ibid., chap. 7.
5. It is said that the actual quote is as follows: "It seems hard to sneak a look at God's cards. But that He plays dice and uses 'telepathic' methods . . . is something that I cannot believe for a single moment." Bill Bryson, *A Short History of Nearly Everything* (New York: Broadway Books, 2003), 146.
6. John Bell, "On the Einstein-Podolsky-Rosen Paradox," *Physics* 1 (1964): 195–200; "On the Problem of Hidden Variables in Quantum Mechanics," *Rev. Mod. Phys.* 38 (1966): 447–52.
7. Anton Zeilinger et al. (letter), *Nature* 445 (Jan. 4, 2007): 65–69.

8. For information on superstrings that is accessible to the general reader, see Michio Kaku and Jennifer Thompson, *Beyond Einstein: The Cosmic Quest for the Theory of the Universe* (New York: Anchor Books, 1995).

9. See Ernest Nagel and James R. Newman, *Gödel's Proof* (New York: New York University Press, 1964).

10. "[T]he doctrine that material objects exist in themselves, apart from the mind's consciousness of them," cited in *Webster's New World College Dictionary*, 3rd ed., s.v. "Realism."

Chapter 3: When Believing Is Seeing

1. Werner Heisenberg, *Physics and Philosophy: The Revolution of Modern Science* (New York: Harper and Row, 1962), 58.

2. Niels Bohr, "Quantum Mechanics and Physical Reality," *Nature* 136 (1935): 65.

3. Cited in Werner Heisenberg, *Physics and Beyond: Encounters and Conversations* (New York: Harper and Row, 1971), 63.

4. George Berkeley, *De Motu*, 1721, in *The Works of George Berkeley*, vol. 4, ed. A. A. Luce and T. E. Jessup (Edinburgh: Nelson, 1952), para. 17.

5. These terms describe a variety of realist philosophies and are not exactly synonymous. *Scientific realists* believe that scientific laws are capable of describing real mechanisms operating behind the veil of appearances. Taking a more radically materialist stance, *scientific materialists* believe that the universe is exclusively material, according to the set of criteria we observed in chapter 1. *Scientism*—more radical still—claims that science is our only source of knowledge about the universe. Similarly, there are subtle differences among the various positions held by instrumentalists, as we will see.

6. Jan Hilgevoord, ed., *Physics and Our World View* (Cambridge, Eng.: Cambridge University Press, 1994), chap. 4.

7. The word has been around for a long time. *Elektron* is the Greek word for amber, which the ancients knew could give you a mild shock if you rubbed something soft against it.

8. Hilgevoord, *Physics and Our World View*, 133.

Chapter 4: The Mind behind the Eyepiece

1. *Webster's New World College Dictionary*, 3rd ed., s.v. "Black box."

2. Cited in Aldous Huxley, *The Perennial Philosophy* (New York: Harper Colophon, 1944), viii.

3. William James, *The Varieties of Religious Experience: A Study in Human Nature* (New York: Penguin, 1985), 388.

4. William James, *Principles of Psychology* (New York: Dover, 1890), 322.

5. From J.B. Watson, *Behaviorism* (New York: People's Institute Publishing Company, 1924), 3.

6. William James, *Essays in Religion and Morality* (Cambridge, Mass.: Harvard University Press, 1989), 85–86.

7. Augustine's four hypotheses are: (1) individual souls derive from those of their parents; (2) they are created from individual conditions at the time of conception; (3) existing elsewhere, souls are sent by God to inhabit human bodies; and (4) souls

descend to the human level by their own choice. Augustine, *The Free Choice of the Will* (Philadelphia: Peter Reilly, 1937), 397.

8. Francisco J. Varela and Natalie Depraz, "Imagining, Embodiment, Phenomenology, and Transformation," in *Buddhism and Science*, ed. B. Alan Wallace (New York: Columbia University Press, 2003), 202.

CHAPTER 5: MASQUERADE

1. *Webster's Ninth New Collegiate Dictionary,* 3rd ed., s.v. "Scientific method."
2. *Webster's New World College Dictionary,* 3rd ed., s.v. "Dogmatism."
3. Richard Feynman, *The Character of Physical Law* (Cambridge, Mass.: MIT Press, 1983), 158.
4. Albert Einstein, *Out of My Later Years* (New York: Philosophical Library, 1950), 59.
5. Bill Moyers, *Healing and the Mind* (New York: Public Affairs Television, Inc., and David Grubin Productions, Inc., 1993).
6. Antonio Damasio, "How the Brain Creates the Mind," *Scientific American,* Dec. 1999, and an updated online version in 2002.
7. *Ibid.*
8. Miguel A. L. Nicolelis and Sidarta Ribeiro, "Seeking the Neural Code," *Scientific American,* Dec. 2006.
9. Steven Pinker, "How to Think About the Mind," *Newsweek,* vol.144, No. 13, Sept. 27, 2004.
10. *The Secret Life of the Brain,* PBS five-part series, 2002.
11. Ibid. Damasio is quoted in the Shatz broadcast.
12. Joel Stein, "Just Say Om," *Time,* Aug. 4, 2003.
13. John Anderson, *Cognitive Psychology and Its Implications* (New York: W. H. Freeman, 1990).
14. Ibid.
15. Damasio, "How the Brain Creates the Mind."
16. *Webster's New World College Dictionary,* 3rd ed., s.v. "Religion."
17. William James, *A Pluralistic Universe* (Cambridge, Mass.: Harvard University Press, 1977), 142.

CHAPTER 6: COMING FULL CIRCLE

1. *Webster's New World College Dictionary,* 3rd ed., s.v. "Consciousness."
2. Francisco J. Varela and Natalie Depraz, "Imagining, Embodiment, Phenomenology, and Transformation," in *Buddhism and Science*, ed. B. Alan Wallace (New York: Columbia University Press, 2003), 200.
3. It has been proposed that the colors perceived by us conform to specific wavelengths of light acting upon receptors in our eyes, but even such a strict one-to-one correspondence has now been disproved.
4. Some may consider this implicit in Christian contemplative traditions, depending on how one interprets terms such as *mind, God, godhead,* and so forth.
5. Originating in India in the second century C.E., the Prasangika Madhyamaka is considered the highest philosophical system by most Tibetan Buddhists.
6. In his *Principles of Psychology,* William James likewise critiqued the atomistic notion of mental events, just around the same time that quantum mechanics was coming on the scene.

7. According to Richard Feynman, "It is important to recognize that in physics today we have no knowledge of what energy *is*." See Richard Feynman, R. B. Leighton, and M. Sands, *The Feynman Lectures on Physics* (Reading, Mass.: Addison-Wesley, 1963).

CHAPTER 7: DECIDING WHAT'S TRUE

1. *Mysticism* is a very general and troublesome term. It is primarily defined as a belief that one can achieve union with the divine through contemplation. However, it is often confused with vagueness and obscurity, as in "I'm mystified." Since contemplative methods are often kept secret or else written in complicated, symbolic language, the two definitions have tended to merge, lending a prejudicial air to the term.
2. Joseph Maréchal, *Studies in the Psychology of the Mystics* (1927; repr., New York: Dover, 2004), 322.
3. Aldous Huxley, *The Perennial Philosophy* (1944; repr., New York: Harper Colophon, 1970), 279.
4. Richard Wilhelm, trans., *The Secret of the Golden Flower* (New York: Harcourt, Brace & World, 1931).
5. This has been aided by Hindu missions to the West, such as the Ramakrishna Mission, whose Vedanta Society of New York was founded in 1898, the Self-Realization Fellowship of Paramahansa Yogananda (1893–1952), and numerous yoga advocates. Indian teachers who achieved popularity in the West, such as Ramana Maharshi (1879–1950), Meher Baba (1894–1969), and Jiddu Krishnamurti (1895–1986) have also given us access to contemplative technology of Hindu origin.
6. B.K.S. Iyengar, *Light on Yoga* (New York: Schocken Books, 1966), 30.
7. Patañjali, *The Yoga-Sutra of Patañjali*, trans. Chip Hartranft (Boston: Shambhala, 2003).
8. *Webster's Ninth New Collegiate Dictionary*, 3rd ed., s.v. "Scientific method."
9. Daniel J. Boorstin, *The Discoverers: A History of Man's Search to Know His World and Himself* (New York: Vintage, 1985).
10. Ian Stevenson, *Where Reincarnation and Biology Intersect* (Westport, Conn.: Praeger Publishers, 1997). Also see Ian Stevenson, *Reincarnation and Biology: A Contribution to the Etiology of Birthmarks and Birth Defects*, 2 vols. (Westport, Conn.: Praeger Publishers, 1997).
11. See Robert G. Jahn, *Margins of Reality: The Role of Consciousness in the Physical World* (New York: Harcourt Brace Jovanovich, 1987).

CHAPTER 8: ILLUMINATING THE BLACK BOXES OF MIND AND MATTER

1. Lao Tzu, *Tao Teh Ching*, trans. John C. H. Wu (Boston: Shambhala, 2005), verse 21.
2. The origin of doctrine of the two truths is found in a teaching given by the Buddha on a mountaintop in northeastern India called Vulture's Peak. There he first expounded on emptiness, and made the statement "Form is emptiness and emptiness is form." Note that "form" is considered relative truth and "emptiness" ultimate truth in this system.
3. This may be what the Buddha had in mind when he said, "Form is emptiness and emptiness is form."
4. Note that in this discussion about the plant being caused by something other than

itself, if we choose to consider causes such as soil, moisture, sunlight, and other contributing causes, this argument will still hold true.

CHAPTER 9: SCIENCE AND CONTEMPLATION

1. Here again we will refer mainly to the explorations of Buddhism due to our familiarity with that tradition, its massive literature on the subject, the availability of that material, and the long and coherent study of the mind that Buddhism has maintained.
2. His Holiness the Dalai Lama, "The Buddhist Concept of Mind," in Daniel Goleman and Robert A. F. Thurman eds., *MindScience: An East-West Dialogue* (Boston: Wisdom Publications), 14.
3. In the case of fMRI, the scientists observe subtle shifts in blood flow in different areas of the brain. With the new, hypersensitive EEG, they are able to pinpoint electrical activity deep within the brain by triangulating from electrodes covering the surface of the head. Another technique, the PET scan (positron emission tomography), employs radioactive tracer dyes to track the activity of neurotransmitters in the brain. (Neurotransmitters are chemicals that act to transmit or inhibit nerve impulses.)
4. A first step in this direction has been taken by the U.S. government's National Institutes of Health with its National Center for Complementary and Alternative Medicine (NCCAM). Established in 1998, NCCAM's mission is to explore complementary and alternative medicine scientifically, train researchers in this area, and disseminate authoritative information.
5. His Holiness the Dalai Lama, *The Universe in a Single Atom* (New York: Morgan Road Books, 2005), 123.
6. See R. M. Ryan, and E. L. Deci, "On Happiness and Human Potentials: A Review of Research on Hedonic and Eudaimonic Well-being," *Annual Review of Psychology* 52 (2001): 141–66; M. Seligman, *Authentic Happiness: Using the New Positive Psychology to Realize Your Potential for Lasting Fulfillment* (New York: Free Press, 2004); and J. Haidt, *The Happiness Hypothesis: Finding Modern Truth in Ancient Wisdom* (New York: Basic Books, 2006).
7. See *Destructive Emotions*, ed. Daniel Goleman (New York: Bantam, 2004).
8. Tim Kasser, *The High Price of Materialism* (Cambridge, Mass.: MIT Press, 2002), 22.

CHAPTER 10: A SCIENCE OF CONSCIOUSNESS?

1. James, *Essays in Religion and Morality*, 87.
2. For a description of the experiences of a Western student of Tibetan Buddhism, Ani Tenzin Palmo, and a glimpse into the world of long-term retreatants, see Vicky MacKenzie, *Cave in the Snow* (New York: Bloomsbury Publishing, 1999).
3. The Great Perfection comprises an advanced and very direct approach to the principal spiritual attainments of Buddhism.
4. His Holiness the Dalai Lama, *Dzogchen, The Heart Essence of the Great Perfection* (Ithaca, N.Y.: Snow Lion Publications, 2000), 172.
5. Located in northeastern India, Nalanda University was probably founded prior to the second century C.E, and lasted until about 1200. An enormous architectural complex, it had some 1,500 teachers and thousands of students studying topics ranging from religion and philosophy to linguistics and medicine.

6. "Rinpoche" is a term of respect and endearment.
7. Henning Genz, *Nothingness: The Science of Empty Space* (Cambridge, Mass.: Perseus, 1999), 26.
8. *Ibid.*, 312.
9. According to the theory of the big bang, between 10^{-35} and 10^{-32} seconds after the big bang, the universe expanded enormously (by a factor of 10^{54}).
10. The term *vacuum* certainly differs in the ways it is conceived by Buddhists of the Great Perfection school and by physicists. But in neither case is it considered in the nihilistic sense of being "merely nothing," an utter negation of content.
11. Albert Einstein, *Out of My Later Years* (New York: Philosophical Library, 1950), 262.
12. Targ, for many years a senior aerospace scientist at Lockheed Martin Corporation, was a pioneer in the development of the laser and later cofounder of a program at Stanford Research Institute that explored psychic abilities. See Russell Targ, *Limitless Mind: A Guide to Remote Viewing and Transformation of Consciousness* (Novato, Calif.: New World Library, 2004).
13. Thomas Aquinas, *Sentences of Peter Lombard*, 4d. 26, 1, 2, cited in Josef Pieper, *Happiness and Contemplation*, trans., Richard and Clara Winston, (Chicago: Henry Regnery Co.,1966), 96.
14. Thomas Aquinas, *Commentary on Aristotle's Nicomachean Ethics* 10, 11; no. 2101, cited in Pieper, 94.
15. His Holiness the Dalai Lama, *The Universe in a Single Atom* (New York: Morgan Road Books, 2005), 208.

PART 3

1. There are occasional exceptions to this sequence. It is said that those with "sharp faculties" may be able to practice the Great Perfection without having to pass through the preliminary stages, but this is considered extremely rare.
2. Those who would like a more detailed introduction to these topics may consult books such as B. Alan Wallace's *Genuine Happiness, The Attention Revolution, Contemplative Science, Hidden Dimensions*, and the Dalai Lama's *Ethics for the New Millennium.*

CHAPTER 11: SHAMATHA

1. The Mahayana school of Buddhism originated in India in the second or first century B.C.E.
2. *Satipatthana Sutta.* See Bhikkhu Bodhi, ed., *In the Buddha's Words* (Somerville, Mass.: Wisdom Publications, Inc., 2005), 291.

CHAPTER 12: VIPASHYANA

1. For an excellent commentary on these practices, see Analayo, *Satipatthana: The Direct Path to Realization* (Birmingham, Ala.: Windhorse Publications, 2003).

CHAPTER 13: THE GREAT PERFECTION

1. Here we encounter a nuance of the term *emptiness*, which was explored earlier. Especially in the Middle Way, one speaks of "emptiness" as referring to the absence of

inherent existence. One example we used was the wristwatch being empty of any inherent wristwatch-ness. We label it a wristwatch, but there is nothing inherent in the collection of its parts that makes it a wristwatch. Here we have made a *simple negation*—inherent existence is negated. However, we can also speak of emptiness as realized indivisibly with primordial consciousness, where it is said to possess a luminous quality manifesting as the display of phenomena. This view is typical of the Great Perfection. In this case a *complex negation* is made—inherent existence is negated, but there is still an experience of primordial consciousness inseparable from the absolute ground of phenomena.

Glossary

antirealism. In the philosophy of science, a general position that questions realist beliefs, especially the claim that the human mind can accurately understand the objective world as it exists independently of all human modes of inquiry. Instrumentalism, pragmatism, and certain forms of empiricism are antirealist philosophies.

Bell's inequalities. The proposition that no physical theory of local hidden variables can ever reproduce all of the predictions of quantum mechanics.

black box. Any phenomenon having a function that can be observed, but whose inner workings are mysterious or unknown.

brain. The physical organ itself, complete with its electrochemical functioning, neuronal pathways, neurotransmitters, and so forth. From this point of view, the mind of subjective experience is considered to be a secondary or emergent phenomenon of the brain. Contrast **mind**.

Buddha-nature. See primordial consciousness.

Chittamatra. Also known as the "Mind Only" school of Buddhism. Views the mind as absolutely real and all other phenomena as illusory, having no reality distinct from the mind.

classical physics. The physics originated by Isaac Newton and the later development of thermodynamics and electromagnetism, which dominated the field until the turn of the twentieth century, when modern physics (including quantum mechanics and relativity) eclipsed many of its underlying assumptions. Classical physics views the universe as a complex machine whose purely objective operations can be understood through scientific measurements.

clear light of death. According to Tibetan Buddhism, the process of dying is comprised of a series of stages or "dissolutions" of the psycho-physical constituents. In the last stage of this process a "clear light" associated with primordial consciousness dawns in awareness.

closure principle. A tenet of scientific materialism that denies the possibility of nonphysical entities affecting anything in the natural world.

common sense. Also known as "conventional wisdom." A collection of shared assumptions accepted without question by a given community.

complementarity. In quantum physics, a general description referring to the dual nature of certain phenomena, including the wave/particle nature of subatomic entities, energy/time in a system, and path/interference aspects of motion.

complementary virtues. A set of principles or criteria used by scientific realists to determine the strength of a scientific theory. They include such characteristics as logical consistency, coherence, consonance with other scientific principles, uniqueness, and fertility.

conservation of energy. A classical scientific law within the field of thermodynamics that states that, within a system, energy can be converted from one form to another, but the total energy within the system remains fixed. Energy can be neither created nor destroyed.

contemplation. A general term for activities that involve engaged attention on a given object. In spiritual traditions, it refers to a variety of practices designed to enhance realization. The term is loosely synonymous with *meditation.*

contemplative insight. (Sanskrit, *vipashyana*) A subtle questioning of the activities and nature of consciousness and the whole world of experience, based upon the attentional stability provided by *shamatha.*

conventional truth. The relative nature of phenomena, which exist as dependently related events and arise in dependence upon (1) prior causes and conditions, (2) their own parts and attributes, and (3) conceptual designation.

definitive meaning. A meaning that is intended to be taken literally.

dharmadhatu. According to Buddhism, the absolute space of phenomena, or the absolute ground for the appearances of external and internal space, time, matter, and consciousness. All phenomena are said to emerge from it.

deduction. Reasoning from the general to the particular. Here a general belief or theory is used to arrive at an understanding of particular details, as when a scientific theory is used to interpret or organize experimental data.

dogma. A coherent, universally applied worldview, consisting of a collection of beliefs and attitudes that call for a person's intellectual and emotional allegiance.

electron. Elementary particle possessing a negative charge, no spatial dimensions, and a mass much smaller than that of the nucleus of an atom. Viewed in terms of the planetary model of the atom, electrons circle the nucleus at fixed distances from it. In conventional terms, the electron is the particle responsible for the flow of electrical current. The Standard Model of subatomic phenomena classifies the electron as a lepton. See the **Standard Model.**

empiricism. A philosophy. In its most basic sense, it is an exploration of phenomena placing emphasis on experience as opposed to theory.

emptiness. According to Buddhist Middle Way philosophy, the absence of inherent existence of all phenomena. Because of their empty nature, all phenomena exist as dependently related events, rather than independently existing things in themselves. See also **Middle Way philosophy.**

entanglement. Also known as "nonlocality." A concept from quantum physics proposing that certain phenomena, such as paired subatomic particles, are intimately interconnected, even at great distances from one another. Includes the notion that the measurement of one of a pair of subatomic particles determines the condition of the two-particle system.

EPR paradox. Named for physicists Albert Einstein, Boris Podolsky, and Nathan Rosen, who proposed that quantum uncertainty was an incorrect theory and that complementary aspects of subatomic entities could be measured. They were proven wrong by Bell's inequalities and subsequent experiments.

functional magnetic resonance imaging (fMRI). A medical technology producing high-quality images of the interior of the body, in real time, using a spectroscopic technique based on the absorption and emission of energy in the radio frequency range of the magnetic spectrum.

hard problem. The scientifically unresolved problem of how the brain generates, or interacts with, subjective experience.

hidden variable. In simple terms, this is an unknown variable in a mathematical formula. The term has wider implications in quantum physics.

induction. A form of reasoning from the particular to the general. Here a theory is arrived at through the study and analysis of experimental data.

instrumentalism. A philosophy declaring that scientific theories should be empirical in nature, based on their practical value for prediction, and that theories should not speculate beyond the data into the realm of metaphysics.

interdependence. A belief from the Buddhist Middle Way (Madhyamaka) philosophy stating that phenomena exist in mutual dependence, rather than having individual, independent, absolute natures.

introspection. The first-person exploration of the mind.

introspectionist school of psychology. A short-lived movement within psychology at the turn of the twentieth century, whose goal was to obtain firsthand knowledge of subjective human states through introspection.

mental perception. The faculty that perceives mental phenomena such as thoughts, emotions, mental images, dreams, and so forth. In Buddhist philosophy it is called the "sixth consciousness," and in this volume is called the "sixth sense."

metaphysical realism. The belief that the universe exists independently of all modes of subjective inquiry and can be known by human beings as it exists in itself.

Middle Way philosophy. Also known as "Madhyamaka" and "Prasangika Madhyamaka." A Buddhist philosophy originating in India in the second century C.E. It seeks a middle position between absolutism and nihilism, stating that phenomena exist interdependently and are therefore empty. See **emptiness.**

mind. A term referring to a broad range of mental phenomena, including our subjective experiences of thoughts, emotions, and perceptions. Contrast **brain.**

nothingness. A nihilist position stating that phenomena are "nothing"—that they are illusory, having no existence whatsoever. Often erroneously confused with the Buddhist concept of emptiness, which accepts the conventional existence of phenomena. See **emptiness.**

nucleus. In physics, within the planetary or solar model of the atom, the central portion composed of protons and neutrons.

objectivism. One of the five tenets of scientific materialism, stating that there is an independent, objective, physical reality outside of our minds and beyond our thoughts, which should be studied without subjective biases.

peer review. A process of verification in science and other fields where proposed ideas and theories are subjected to the (usually public) inspection and criticism of one's professional peers.

Perfection, Great.   A Buddhist system of theory and practice traced back to the Indian contemplative Prahevajra, (born according to traditional accounts 360 years after the Buddha's death), and now preserved primarily in the Nyingma order of Tibetan Buddhism; its central concern is the realization of primordial consciousness.

person.   Used here in the following manner: *First-person* information is that obtained directly by the subject, as in a subject's observation of his or her own thoughts. *Second-person* information is obtained by an expert when interviewing a subject. *Third-person* information comes from the examination of data, as when a scientist observes test results.

photon.   Light viewed as particles, or electromagnetic energy packets.

physical reductionism.   A principle of scientific materialism stating that nature can be reduced to physical entities and their functions, each isolated from the rest, having no connections save the patterns imposed by the laws of nature.

physics of the vacuum.   A branch of science focusing on the nature of vacuums and having applications to the origins of the physical universe and the big bang.

pluralism.   A philosophy of science suggesting that scientific fields should develop their own views and procedures based on their particular subject matter, rather than conforming to the traditional pecking order that begins with physics (as the "cutting edge" science) and continues on down to chemistry, biology, and so forth.

pragmatism.   A philosophy placing emphasis on specifics ("small narratives") rather than widely embracing theories ("grand narratives"). In the philosophy of science, pragmatism is a radical form of antirealism, demanding that theories not reach beyond the practical applications of experimental data.

primordial consciousness.   (Sanskrit, *jnana*) In Buddhism the ultimate ground state of consciousness, out of which all appearances emerge and into which they subside.

provisional meaning.   A meaning that is not to be taken literally, but that is open to interpretation relative to specific contexts.

psyche.   Dualistic conscious and subconscious mental processes, including the five physical senses, arising moment to moment as a coarse mind stream. This is what psychologists study and the corresponding term in Buddhism is "coarse mind."

quantum physics.   A branch of modern physics, established circa 1900, focusing primarily on subatomic entities. Its conclusions, which contradict many of the tenets of classical physics, suggest a subatomic environment that is "fuzzy" and indeterminate—where absolutely "objective" observations and measurements are not possible. Its main tenets include the wave/particle and other dualities, superposition, uncertainty, and entanglement (nonlocality).

realism.   The belief that material objects exist in themselves, apart from the mind's consciousness of them.

reincarnation.   The belief that after their deaths living beings are reborn. Religions that share this belief include Hinduism and Buddhism, among others.

relativity.   The scientific view, based on theories presented by Albert Einstein at the beginning of the twentieth century, that time and space are not absolute, as they are in classical physics. The physical proportions of such features as velocity, matter, and energy change according to context. The absolutes of space and time of classical physics are replaced by space-time, where these two are mutually influential. Gravity is viewed as a feature of the curvature of space-time.

Sautrantika.   A Buddhist philosophy claiming that both physical and mental phenomena are "concrete," that is, they are real and exist independently of any conceptual framework, although they are constantly changing and engaging one another.

Scientific materialism.   An interpretation of science wherein only physical phenomena are considered real. It is based on five tenets: objectivism, metaphysical realism, the closure principle, universalism, and physical reductionism.

scientific method.   The principles and procedures for the systematic pursuit of knowledge, involving the recognition and formulation of a problem, the collection of data through observation and experiment, and the formulation and testing of hypotheses.

scientific realism.   The belief that science can describe reality as it exists independently of any observer.

scientism.   The belief that science is the only source of knowledge of the universe.

Standard Model.   A view of subatomic entities, developed beginning in the 1930s, that incorporates quantum physics. This model envisions the subatomic world as divided into two major types of particles and four forces (gravity, electromagnetism, the strong nuclear force, and the weak nuclear force). The Standard Model is considered an advance over the previous, solar model of the atom.

shamatha. Meditative quiescence. A term for a range of meditative techniques in Buddhism, with parallels in other contemplative traditions, where the progressive development of attention brings one to an exceptional state of relaxation, stability, and vividness.

substrate. (Sanskrit, *alaya*) In Buddhism, the objective, empty space of the individual's mind into which all objective appearances of the physical senses and mental activity dissolve when one falls asleep, when one achieves shamatha, and when one dies.

substrate consciousness. (Sanskrit, *alayavijnana*) As explained in Buddhism, the subtle mind stream of the individual sentient being maintained throughout all of its incarnations. All of the latent and subconscious phenomena that we commonly attribute to the mind and the senses are said to emerge from the substrate consciousness.

superposition. Term from quantum physics holding that subatomic phenomena in motion actually exist as a mixture of chance possibilities until they are measured.

superstring theory. A branch of physics, developed beginning in the 1980s, that views elementary particles as one-dimensional, vibrating, string-like entities. The theory aims to unite both quantum mechanics and the gravity of curved space-time into a single, homogeneous view of the universe.

tantra. Term generally applied to the varied practices of Vajrayana Buddhism.

thermodynamics. The branch of physics that studies the effects of changes in temperature, pressure, and volume on physical systems at the macroscopic scale by analyzing via statistics the collective motion of their particles.

two truths. A Buddhist philosophy proposing two levels of truth: conventional and ultimate. Briefly, conventional truths do not exist in the way they appear, while ultimate truths do. See **conventional truth** and **ultimate truth**.

ultimate truth. The absence of inherent nature in phenomena, existing independently from any conceptual framework. See **emptiness** and **two truths**.

uncertainty. A principle of quantum physics deriving from the problem of measurement in the subatomic environment. Since all of the measurements needed to satisfactorily define subatomic phenomena according to classical physics are not obtainable, these attributes (position and momentum, the initial state of a system, and so on) are uncertain.

underdetermination. A term from the philosophy of science applied to situations where multiple theories are equally effective in accounting for the same experimental data. In such cases an objective choice as to which theory is "correct" of "better" cannot be determined.

universalism.   A tenet of scientific materialism stating that scientific laws, especially the laws of classical physics, apply equally in every part of the universe.

Vaibhashika.   A Buddhist atomist philosophy, stating that tiny, irreducible atoms are the only physical phenomena having absolute existence and that brief, irreducible moments of consciousness are the absolute level of the mind.

vacuum, absolute.   In physics, it is called "true vacuum" and refers to all that remains after removing everything that the laws of nature permit us to remove from a volume of space. Contrast with relative vacuum.

vacuum, relative.   In physics, it is called "false vacuum" and refers to the state of a system that is not the lowest possible energy state but is nonetheless stable for some time. Contrast with absolute vacuum.

vipashyana.   See **contemplative insight.**

wave/particle duality.   Concept from quantum physics describing the dual nature of light and other subatomic entities. Depending on the form of measurement or observation, such phenomena can be viewed as either waves or particles.

working hypothesis.   A form of scientific inquiry wherein theories are presented as mere proposals whose development and proof depends on confirmation from experimental evidence.

Bibliography

Analayo. *Satipatthana: The Direct Path to Realization*. Birmingham, Ala.: Wind-horse Publications, 2004.

Anderson, John. *Cognitive Psychology and Its Implications*. New York: Worth Publishers, 1990.

Augustine. *The Free Choice of the Will*, trans. Francis E. Tourscher. Philadelphia: Peter Reilly, 1937.

Berkeley, George. *De Motu*, 1721. In *The Works of George Berkeley*, edited by A. A. Luce and T. E. Jessup. Edinburgh: Nelson, 1952.

Bodhi, Bhikkhu, ed. *In the Buddha's Words*. Somerville, Mass.: Wisdom Publications, Inc., 2005.

Boorstin, Daniel J. *The Discoverers: A History of Man's Search to Know His World and Himself*. New York: Vintage, 1985.

Bryson, Bill. *A Short History of Nearly Everything*. New York: Broadway Books, 2003.

His Holiness the Dalai Lama. *Dzogchen, The Heart Essence of the Great Perfection*. Ithaca, N.Y.: Snow Lion Publications, 2000.

———. *The Universe in a Single Atom*. New York: Morgan Road Books, 2005.

_____ et al. Daniel Goldman and Robert A.F. Thurman, eds. *MindScience: An East-West Dialogue*. Cambridge, Mass.: Wisdom Publications, 1991.

Einstein, Albert. *Out of My Later Years*. New York: Philosophical Library, 1950.

Feynman, Richard. *The Character of Physical Law*. Cambridge, Mass.: MIT Press, 1983.

Feynman, Richard P., R. B. Leighton, and M. Sands. *The Feynman Lectures on Physics*. Reading, Mass.: Addison-Wesley, 1963.

Galileo, *Il saggiatore (The Assayer)*, 1623. In *Discoveries and Opinions of Galileo*, trans. Stillman Drake. New York: Knopf, 1957.

Genz, Henning. *Nothingness: The Science of Empty Space*. Cambridge, Mass.: Perseus, 1999.

Goleman, Daniel, ed. *Destructive Emotions*. New York: Bantam, 2004.

Greenstein, George, and Arthur G. Zajonc. *The Quantum Challenge.* Sudbury, Mass.: Jones and Bartlett, 1997.

Heisenberg, Werner. *Physics and Beyond: Encounters and Conversations.* New York: Harper and Row, 1971.

————. *Physics and Philosophy: The Revolution in Modern Science.* New York: Harper and Row, 1962.

Hilgevoord, Jan, ed. *Physics and Our World View.* Cambridge, Eng.: Cambridge University Press, 1994.

Huxley, Aldous. *The Perennial Philosophy.* 1944. Reprint, New York: Harper Colophon, 1970.

Iyengar, B. K. S. *Light on Yoga.* New York: Schocken Books, 1966.

Jahn, Robert G. *Margins of Reality: The Role of Consciousness in the Physical World.* New York: Harcourt Brace Jovanovich, 1987.

James, William. *Essays in Religion and Morality.* Cambridge, Mass.: Harvard University Press, 1989.

————. *A Pluralistic Universe,* Cambridge, Mass.: Harvard University Press, 1977.

————. *Principles of Psychology.* 1890. Reprint, New York: Dover, 1950.

————. *The Varieties of Religious Experience: A Study in Human Nature.* 1902. Reprint, New York: Penguin, 1985.

Kasser, Tim. *The High Price of Materialism.* Cambridge, Mass.: MIT Press, 2002.

Koestler, Arthur. *The Ghost in the Machine.* New York: Macmillan, 1967.

Lao Tzu. *Tao Teh Ching.* Translated by John C. H. Wu. Boston: Shambhala, 2005.

MacKenzie, Vicky. *Cave in the Snow.* New York: Bloomsbury Publishing, 1999.

Maréchal, Joseph. *Studies in the Psychology of the Mystics.* 1927. Reprint, New York: Dover, 2004.

Nagel, Ernest, and James R. Newman. *Gödel's Proof.* New York: New York University Press, 1964.

Patañjali. *The Yoga-Sutra of Patañjali.* Translated by Chip Hartranft. Boston: Shambhala, 2003.

Pieper, Josef. *Happiness and Contemplation.* Translated by Richard and Clara Winston. Chicago: Henry Regnery Co., 1996.

Stevenson, Ian. *Reincarnation and Biology: A Contribution to the Etiology of Birthmarks and Birth Defects.* Westport, Conn.: Praeger Publishers, 1997.

————. *Where Reincarnation and Biology Intersect.* Westport, Conn.: Praeger Publishers, 1997.

Targ, Russell. *Limitless Mind: A Guide to Remote Viewing and Transformation of Consciousness.* Novato, Calif.: New World Library, 2004.

Thoreau, Henry David. *Walden.* New York: W. W. Norton & Company, Inc., 1951.

Wallace, B. Alan, ed. *Buddhism and Science*. New York: Columbia University Press, 2003.

_____. *Contemplative Science: Where Buddhism and Neuroscience Converge*. New York: Columbia University Press, 2006.

Watson, J. B. *Behaviorism*. New York: People's Institute Publishing Company, 1924.

Wilhelm, Richard, trans. *The Secret of the Golden Flower*. New York: Harcourt, Brace & World, 1931.

Index